ASTRONOMY

ASTRONOMY

IAN RIDPATH

GALLERY BOOKS
An Imprint of W. H. Smith Publishers Inc.
112 Madison Avenue
New York City 10016

This edition first published in 1991 by Gallery Books, an imprint of W. H. Smith Publishers, Inc., 112 Madison Avenue, New York, New York 10016

Published in England by Dragon's World Ltd, Limpsfield and London

Editorial: John Woodruff, Diana Steedman
Editorial Director: Pippa Rubinstein
Design: David Hunter

Gallery Books are available for bulk purchase for sales promotion and premium use. For details write or telephone the Manager of Special Sales, W. H. Smith Publishers, Inc., 112 Madison Avenue, New York, New York 10016. (212) 532-6600.

ISBN 0·8317·0476·4

Printed in Singapore

CONTENTS

1 STARGAZERS

Come with me, back to a time when the night sky was truly dark, untainted by artificial lights or the smoke of civilization. We shall travel to the Mediterranean, where ancient peoples first traced out the constellation patterns and gave them names that live on today.

I have with me a young companion. She is dressed in the light tunic of a caryatid, fastened with a brooch at the shoulder and gathered at the waist with a string.

She points upward, to a star nearly overhead – a sapphire in the soft velvet of night. "The bright star on the shell," she calls it. (We know it as Vega, an Arabic name originating a millennium after the Greeks.)

My companion holds a book, a poem about the sky written by Aratus, poet laureate to King Antigonus of Macedonia in northern Greece around 275 BC. She quotes his description of Lyra: "The tiny tortoise which, while still beside his cradle, Hermes pierced for strings and bade it be called the lyre."

Aratus's poem, entitled the *Phaenomena* ('appearances'), is the earliest description we have of the constellations that were known to the ancient Greeks. To them, a constellation was a pattern of stars that suggested a character or object from their mythology – the lyre made from a tortoise-shell by the infant Hermes, the tableau of Perseus and Andromeda, the exploits of Zeus, and so on.

Four centuries after Aratus came the last of the great Greek astronomers, Ptolemy, who listed 48 constellations in his synthesis of Greek astronomical knowledge, the *Almagest*. (This is the title medieval Arab astronomers respectfully gave it – 'the greatest'.) Ptolemy's original 48 constellations survive to this day, augmented by 40 others added since the sixteenth century to fill gaps, particularly in the southern skies which were forever below the Mediterranean horizon.

My companion hesitates, as though she is afraid of asking the unanswerable. Then she poses the question that must have been asked since the dawn of time: "What are the stars?"

"They are glowing balls of gas – other Suns, but vastly more remote," I answer, with the benefit of two thousand years of hindsight.

She nods, and I see that the idea is not totally strange to her. "What about the wandering stars, the planets?"

● The Starry Night *by the Dutch artist Vincent van Gogh, painted in June 1889 in the south of France, shows a golden crescent Moon against a sky shimmering with bright stars and planets. The swirling pattern in the sky may have been inspired by Lord Rosse's drawing of the Whirlpool Galaxy (see p. 145). Night sky scenes featured in several of van Gogh's paintings, and this is the most famous of them.*

Oil on canvas, 73.7 × 92.1 cm. Collection, the Museum of Modern Art, New York. Acquired through the Lillie P. Bliss Bequest.

"They are other worlds in space. They go around the Sun, like the Earth does. But there is no life on them." At this she is quiet, for the idea is an immense one.

She looks again at the sky. In the enclosing darkness, the Milky Way arches overhead like a phosphorescent skein of wool. "That is a mass of stars," I tell her, "even more distant than those that make up the constellations."

She gasps in delight as a shooting star blazes across the firmament.

"A piece of dust from space, burning up in the atmosphere above us," I explain. It is difficult to know how far she comprehends the idea of an atmosphere or surrounding space.

"And what makes the stars shine?" she asks finally.

Here I am at a loss. How can I tell her about nuclear fusion, the science of the the twentieth century that ordinary people even today cannot understand?

"Energy is given out deep inside them by atoms being crushed together," I reply. The Greeks were familiar with the idea of atoms, as developed by Democritus around 400 BC.

When the Greeks invented the constellations – and they are indeed merely inventions of the human imagination, with no physical reality – they did not realize that the stars are scattered at widely differing distances from us. At that time the stars were thought to lie on a transparent sphere of indeterminate size, although there was a general feeling that they were substantially more distant than the Sun.

In the seventeenth century, when Galileo's observations and Johann Kepler's calculations finally established that the Earth is a planet orbiting the Sun, the proposition that stars are other suns became accepted, and a number of attempts were made to estimate their distances by assuming that they were of similar brightness to the Sun. On this basis Isaac Newton estimated that Sirius is about a million times as far away as the Sun, twice the real value but nevertheless a useful guide.

Not until 1838 did the first accurate measurement of a star's distance become available. In that year the German astronomer Friedrich Wilhelm Bessel detected a minuscule change in position of the star 61 Cygni as observed from different points along the Earth's orbit. From such a position change, termed a *parallax*, can be calculated the star's distance (see diagram). Bessel's announcement was rapidly followed by determinations of the parallax of two other stars, Alpha Centauri and Vega.

These observations confirmed that the stars are so far away that their light takes many years to reach us. Thus was born the *light year*, a unit of distance (*not* time) equal to how far a beam of light travels in one calendar year. Light travels at the fastest speed in the Universe, 300,000 km/sec (186,000 mile/sec), and so a light year is equivalent to nearly 9.5 million million km (5.9 million million miles). A related unit of distance is the *parsec*, the range at which a star would have a parallax of one second of arc ($\frac{1}{3600}$ of a degree), and equal to just over three-and-a-quarter light years.

● *A marble statue of Atlas carrying on his shoulders not the world, but a celestial globe on which are engraved the constellations as they were visualized by the ancient Greeks. This statue, known as the Farnese Atlas, was made in the second century AD and is believed to be a copy of an earlier original from the time of the Greek poet Aratus, when the constellations as we know them were first laid down.*

● Right: *Star motions gradually change the shape of the constellations. Here are the main stars in the constellations of Ursa Major and Leo as they appeared 100,000 years ago, how they appear today, and how they will appear 100,000 years hence.*

HOW STARS GET THEIR NAMES

The brightest stars in each constellation are identified by Greek letters – α (alpha), β (beta), and so on – a system begun by the German celestial cartographer Johann Bayer in 1603. Other stars not covered by this system are referred to by letters or numbers allocated to them in star catalogues. In addition, some stars bear names, usually Greek, Roman, or Arabic in origin. Letter or number designations are used with the genitive of the constellation's name: for example, α Orionis means 'alpha of Orion', and 61 Cygni means '61 of Cygnus'.

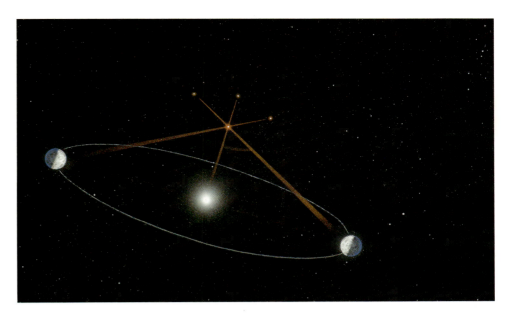

● Above: *The German astronomer Friedrich Wilhelm Bessel was the first to measure the parallax of a star, 61 Cygni. Later he found companion stars to Sirius and Procyon.*

● Above right: *Distances of nearby stars can be found by measuring their change in position as seen from opposite sides of the Earth's orbit. A star's displacement from its average position is known as its parallax.*

One thing that would have surprised the Greeks is the knowledge that the constellation patterns are very slowly changing. This was first realized in 1718 by Edmond Halley, when he noted that several bright stars – including Sirius and Arcturus – had changed position perceptibly since the time of the ancient Greeks. We now know that all stars have such *proper motions*, to a greater or lesser degree. Over time, proper motions distort the shapes of the constellations so that eventually the familiar figures that were framed by the ancient Greeks will become unrecognizable.

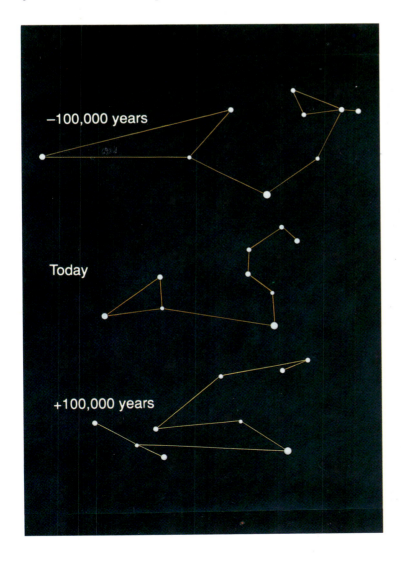

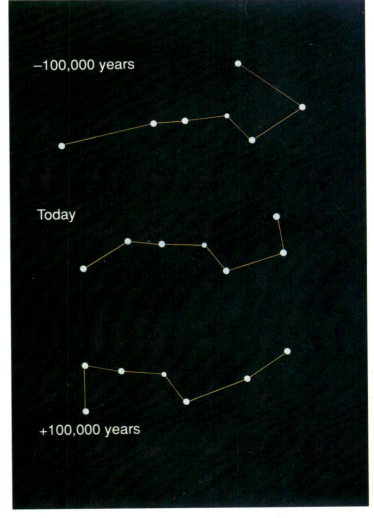

POLLUTING THE NIGHT

We live in an age that has touched the Moon but, paradoxically, is cutting itself off from the Universe of which we are a part. The night sky has been polluted by artificial light to such an extent that in urban areas it is difficult to recognize even the brightest stars and constellations. Most people today know less about the sky than did the ancient Greeks 2000 years ago.

Pictures such as this demonstrate how much light is leaking wastefully into space. On this whole-Earth photograph, a montage of night-time images from US Air Force weather satellites, major conurbations can easily be identified and many countries are outlined by the lights from coastal towns. Enlargements of the USA and Europe are shown below.

There is much fascinating detail to be seen. Some countries, France for example, seem quite dark, while others such as Japan are swamped with artificial light. The Nile valley is brightly outlined; Puerto Rico blazes in the Caribbean; Moscow is surrounded by a spokelike pattern. Transportation features such as the Trans-Siberian railway and Interstate Highway 5 along the western USA can be picked out. The large blob west of Japan comes from a fishing fleet using brilliant lights to lure squid to the surface.

The brightest features are not artificial lights as such, but mark the burning of natural gas from oil wells, most notably in the Middle East, Siberia, and north Africa. Tiny specks of light across central Africa and southeast Asia show where grassland and forests are being burnt for agricultural purposes. The swirl of light over Greenland and northern Canada is an aurora.

Among the questions raised by these images is why there should be such a contrast between the USA, Europe, and Japan, where one-quarter of the world's population uses three-quarters of the world's electricity, and the darkness of Asia, Africa, and South America. Why is so much street lighting being wasted into space? Better lighting practices would save energy, reduce unwanted emission of greenhouse gases from power stations, and preserve the night sky.

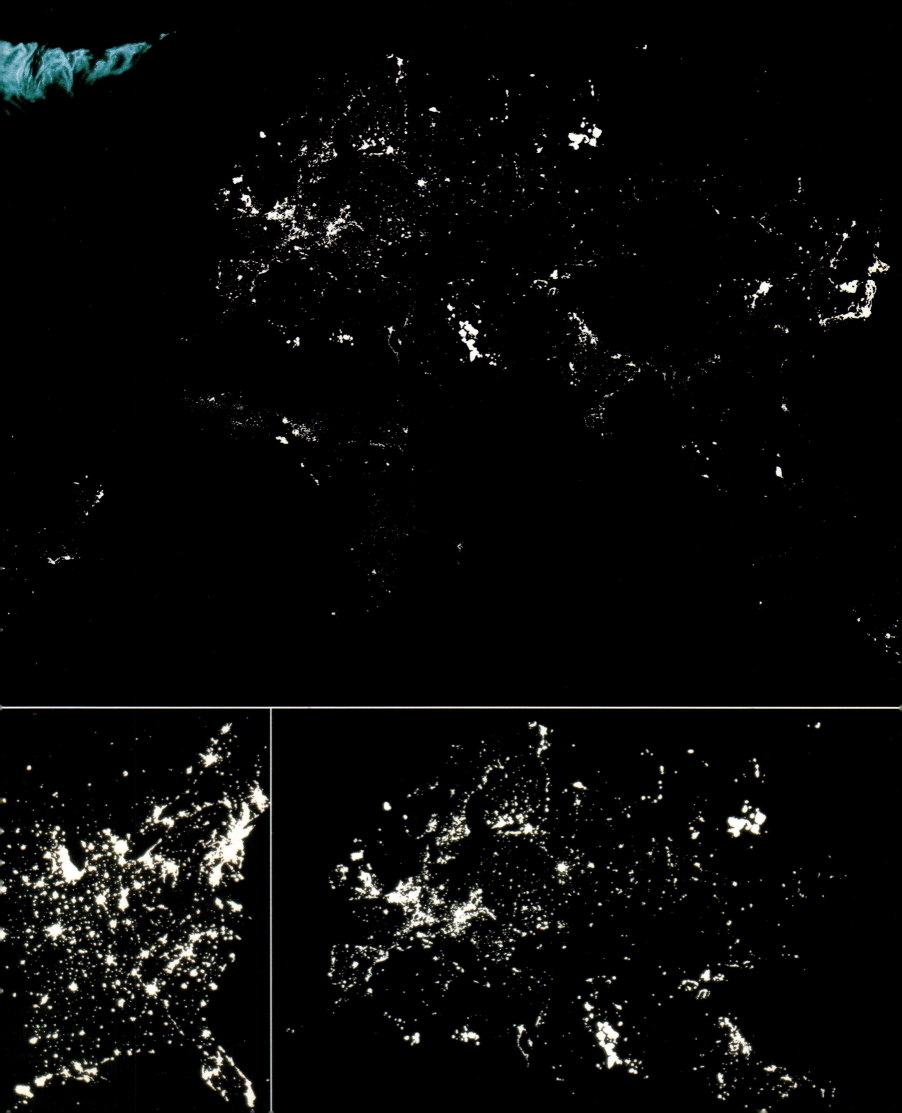

"If you look with care you will see that certain stars are lemon-coloured, others have pink, green, blue, or myosotis glints. In painting a starry sky it is clearly not enough to put white specks on blue–black." So wrote the artist Vincent van Gogh, who produced some of the best-known paintings of the night sky. The colours he referred to, which are more prominent in binoculars and telescopes than they are to the naked eye, are of more than artistic interest. They are a guide to the nature of each star, for the colour tells us the star's temperature and allows us to deduce much else besides.

Do not, incidentally, be fooled by stars close to the horizon that appear to flash a multitude of colours. This is nothing to do with the star itself, but is caused by turbulence in the atmosphere; the nearer a star is to the horizon, the more of the atmosphere its light has to pass through to reach us, and the more it is affected by turbulence. Sirius, the brightest star in the night sky, is particularly prone to such colourful scintillation when it is close to the horizon. Swimming air currents break up the star's light into all the colours of the rainbow so that it sparkles a multitude of colours from red to blue. The true colour of Sirius is a bluish white.

The coolest stars glow the reddest, while progressively hotter stars are yellow and white; the most intensely hot stars appear blue–white.

● *Old star maps, such as this one from 1660 by the Dutchman Andreas Cellarius, lovingly depicted the mythological characters of the Greeks. But the fashion for pictorial maps died out in the nineteenth century as star-plotting grew more refined and sky maps became the province of the surveyor rather than the mural painter. Nowadays a constellation is regarded simply as an area of sky rather than a picture, although the names of the heroes and villains of celestial myth live on.*

● *The Sun compared in size with larger and smaller stars, from a supergiant to a white dwarf.*

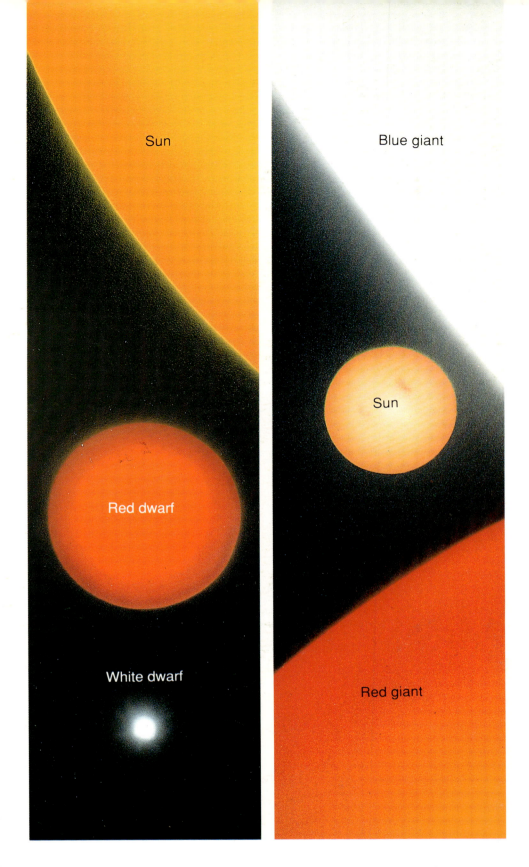

Sun

Blue giant

Sun

Red dwarf

Red giant

White dwarf

HOW ASTRONOMERS MEASURE STAR SIZES

Since stars appear as nothing more than points of light in even the most powerful Earth-based telescopes, how can astronomers measure their sizes? It has to be done indirectly, from the temperature and luminosity (the 'wattage', in everyday terms) of a star. The temperature is easy enough to determine from the star's colour, but to estimate the luminosity we must first know its distance, and that is much more difficult to pin down. Distances are fairly well established for the nearest stars, and from these we can estimate the distances to similar stars that are farther away.

Once we have established a star's luminosity and temperature, it is possible to work out its total surface area and hence its diameter.

Arcturus, Antares, and Betelgeuse, for example, are noticeably orange or reddish, whereas Rigel is distinctly bluish. All four stars are of the types called *giants* and *supergiants*, whose distinguishing characteristics are their bloated size and extreme brilliance.

Arcturus, a red giant in the constellation Boötes, the herdsman, is about twenty-five times the diameter of the Sun. Yet Arcturus is unexceptional when placed alongside supergiants such as Betelgeuse and Antares, which are ten times larger still – so large that, if placed where the Sun is now, they would not only engulf the Earth, but reach beyond the orbit of Mars. A beam of light, or a radio signal, would take about an hour to circumnavigate such gargantuan stars. Among the most distended of all supergiants is Mu Cephei, which has an estimated diameter over 3000 times that of the Sun, enough to swallow the orbit of Saturn with plenty of room to spare.

Red giants and supergiants are much more luminous than the Sun because they have greater surface areas, even though their surfaces are cooler. Arcturus, for example, has a surface temperature of 4200°C, against 5500°C for the Sun, which is a yellow–white star, but still Arcturus gives out the power of 100 Suns. Betelgeuse, a supergiant, is 100 times as luminous as Arcturus, even though it is cooler, with a surface temperature of about 3000°C.

Two factors govern the amount of light that a star gives out: its diameter and its temperature. A blue giant or supergiant is vastly more luminous than a red star of the same diameter because the blue star is much hotter. Actually there are no blue stars as large as Betelgeuse, but the familiar blue–white supergiant Rigel gives out twice as much light as Betelgeuse even though it is only about one-tenth the diameter. Rigel's temperature, 10,000°C, is over three times Betelgeuse's, but still only about a quarter of that of the hottest-known stars such as the blue supergiant Zeta Puppis.

● Right: *Star trails around the north celestial pole, over the dome of the William Herschel telescope on La Palma in the Canary Islands. The trails are produced by the rotation of the Earth during a time exposure. The short, bright arc near the right edge is the trail of Polaris, which lies about a degree from the true north celestial pole.*

● Below: *A galaxy of stars, NGC 253 in the constellation Sculptor. It is spiral in shape but is tilted at an angle to us. Stars form within gas and dust clouds in the spiral arms of galaxies. Our Sun is one star in a galaxy such as this. Other galaxies are dotted throughout the Universe as far as telescopes can see.*

At the other end of the scale are the dwarf stars, some only a few per cent of the Sun's diameter. These stellar midgets are so faint that none can be seen with the naked eye, not even Proxima Centauri, the closest of all stars to the Sun. Proxima Centauri and another solar neighbour, Barnard's Star, are *red dwarfs*, similar in surface temperature to Betelgeuse but having only about one-tenth the diameter of the Sun. Red dwarfs are veritable glow-worms, emitting just one ten-thousandth as much light as our Sun. If such a star were substituted for the Sun, the Earth would be bathed in an angry twilight by day, and the full Moon would be a pale red disk at night, incapable of throwing shadows.

But even red dwarfs are not the smallest inhabitants of the stellar menagerie. A further step down the scale are the *white dwarfs*, one-tenth the size again but with a similar light output because of their higher surface temperature. A typical white dwarf is the size of the Earth, but contains as much matter as the Sun squeezed to a density far greater than the heaviest metal on Earth. White dwarfs are the end points in the lives of stars such as the Sun.

Sirius, the brightest star in the night sky, and Procyon, another brilliant winter star, both have white dwarf companions which orbit them in 50 and 40 years, respectively. But these white dwarfs are very difficult to see even in a large telescope, because they are swamped by the brilliance of the nearby parent. The easiest white dwarf to see is a companion of a star called Omicron-2 Eridani (also known as 40 Eridani). The main star is similar to our Sun, and can be seen with the naked eye. A small telescope reveals the white dwarf, and there is an even fainter red dwarf companion as well.

Stars considerably more massive than the Sun form neutron stars or black holes when they die, as will be explained in Chapter Two, where we shall see how all these different types of star fit into a coherent pattern of evolution.

The underlying factor that governs the size and brightness of a star is its mass – the amount of gas it contains. There is an upper mass for a star, around 100 Suns, above which the star will be so hot and luminous that it will literally puff itself apart. Among the most massive known stars is the supergiant Eta Carinae (see p. 31), a powerhouse 5 million times as energetic as our Sun. There is also a minimum mass for a star, about 6 or 7 per cent of the Sun's mass, below which the temperature and pressure at the centre never reach the critical values necessary to spark off the nuclear reactions that keep a true star glowing.

In recent years astronomers have begun to find objects that lie just on or below the minimum mass for a true star. These objects are called *brown dwarfs*. They are even cooler than red dwarfs and have diameters similar to that of the planet Jupiter, although they are considerably more massive than Jupiter. By their nature they are difficult to detect and the only reason they shine at all is through the heat left over from their formation. Brown dwarfs are 'in between' objects that bridge the gap between planets and stars. One day it may be possible to detect objects with even smaller masses than brown dwarfs, one per cent of the Sun's mass (ten times Jupiter's) or less. Anything this small would be considered a planet.

● Left: *Stars form inside clouds of hydrogen gas such as this one, the Cone Nebula in the constellation Monoceros, which takes its name from a dark cone-shaped wedge of dust reaching across the brighter background like an elephant's trunk. The Cone Nebula is over 2500 light years away, and the dark dust cloud is well over 5 light years from end to end.*

● Following page: *An immense ball of stars, the globular cluster 47 Tucanae shimmers at the edge of our Galaxy, 13,000 light years from us. It is visible to the naked eye from southern skies as a faint smudge, but long-exposure photographs like this one taken through large telescopes reveal the 100,000 or so individual stars, many of them larger and brighter than our own Sun, that comprise the cluster.*

Until the invention of the telescope early in the seventeenth century, astronomers had only their eyes with which to survey the sky. Telescopes created a revolution in astronomy because they not only showed known objects such as the Moon and planets in more detail, they also revealed countless new stars whose existence had never been suspected. Under telescopic scrutiny a suspiciously large number of faint stars turned out to lie cheek-by-jowl with known ones. There were simply too many pairings for them all to be coincidental alignments of stars at different distances, and the inevitable conclusion was that in most cases the stars were genuinely related.

The British astronomer Sir William Herschel discovered hundreds of double stars and carefully noted the relative positions of each member of a pair. Over the decades he re-observed each double. In six cases he found definite movement of one star in orbit around the other, thereby proving that they are genuine twins, which astronomers term *binaries*. His most famous example was the bright star Castor in Gemini. Telescopes show that it consists of two blue–white stars, which Herschel estimated orbit each other in about 350 years; the true figure is just over 450 years. But there is more to Castor than this. There is a fainter red dwarf nearby, which is related to the other two, and detailed study of the light from all three stars has revealed that each has a close twin unseen in any telescope. Castor is therefore a family of six stars linked by gravity.

Starlight is analysed by an instrument called a spectroscope (see p. 43), and double stars that are revealed in this way are termed *spectroscopic binaries*. Spectroscopic binaries usually have an orbital period of a few days, and some actually eclipse each other as they orbit, as we shall see later. By contrast, double stars that are wide enough to be distinguished, or 'separated', through telescopes have orbital periods that range from decades to centuries and even longer.

Since Herschel's time astronomers have come to recognize that the great majority of stars are double or multiple, as though nature's fecundity discourages the production of single stars. In this respect, therefore, our Sun is unusual, although in place of stellar companions it has planets.

STAR BRIGHTNESSES

Astronomers measure the brightnesses of stars in terms of magnitudes, a system first used by ancient Greek astronomers who classified the brightest stars as first magnitude and the faintest visible to the naked eye as sixth magnitude. Nowadays astronomers define a star of sixth magnitude as exactly 100 times fainter than one of first magnitude. Stars fainter than sixth magnitude are assigned progressively to seventh magnitude, eighth magnitude, and so on. The faintest objects that can be seen through Earth-based telescopes are of about 25th magnitude; instruments such as the Hubble Space Telescope in orbit around the Earth extend this limit to around 30th magnitude, 100 times fainter.

For brighter objects, the magnitude scale is extended indefinitely in the opposite direction, through magnitude zero to negative magnitudes. Sirius, the brightest star in the sky, is magnitude -1.46. The planets Jupiter and Venus are magnitude -2.9 and -4.7, respectively, at their brightest, and under dark skies Venus can even cast shadows. The full Moon is magnitude -12.7 and the Sun -26.8.

These are all *apparent magnitudes*, i.e. as they appear to us from Earth. More important for the purposes of comparing stars with each other is their *absolute magnitude*, which is how bright they would appear if they were all a standard distance away. (This standard distance is set by convention at 10 parsecs.) The absolute magnitude of the Sun is a modest $+4.8$, which means it would not be at all remarkable to the naked eye if seen from 10 parsecs away. At the same distance Sirius would be magnitude $+1.4$, whereas Betelgeuse and Rigel would be about magnitudes -7 and -8 respectively. The most luminous stars known have absolute magnitudes of about -10, whereas the absolute magnitudes of the faintest red dwarfs are about $+18$.

Another celebrated example in the history of double stars is the second star along the handle of the Plough, called Mizar. People with sharp eyes will see a fainter companion, known as Alcor, which is unmistakable through binoculars. Mizar and Alcor are too far apart to form a true binary; they move through space together but do not seem to orbit each other. However, Mizar has another companion, closer to it than Alcor and visible through a telescope. This closer star was first seen in 1650 by the Italian astronomer Giovanni Riccioli, making Mizar the first double star to be discovered. In 1889, through the work of the American astronomer E. C. Pickering, Mizar became the first spectroscopic binary to be detected. Subsequently, astronomers have established that Mizar's closer companion and Alcor are spectroscopic binaries too.

The closest star to the Sun is a multiple system. To the naked eye, Alpha Centauri appears as a brilliant yellow–white star. Telescopes divide it into twin Sun-like stars that orbit each other in 80 years. The third member of the system, the red dwarf called Proxima Centauri, lies about one-tenth of a light year closer to us than the bright pair, and is believed to take as much as a million years to orbit them.

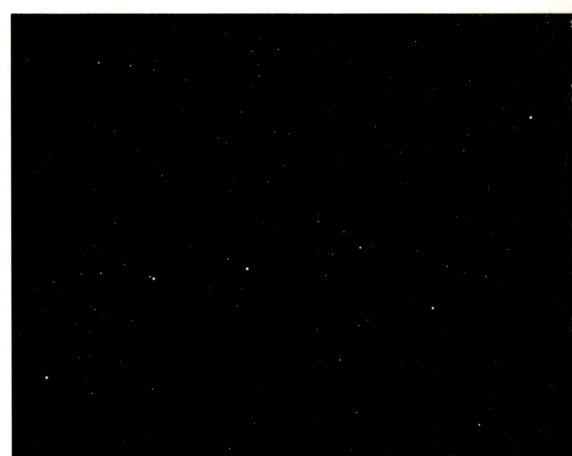

● Above: *The familiar shape of the Plough, or Big Dipper, consisting of the seven brightest stars in the constellation Ursa Major, the Great Bear. The second star along the handle, Mizar, is a wide double. Its companion star, Alcor, is visible with sharp eyesight.*

● Left: *A double star, U Geminorum, consisting of a white dwarf with a larger orange companion, as it might appear from a hypothetical planet. Gas from the companion produces a glowing ring around the white dwarf. Every few weeks or months this ring of gas suddenly increases in brightness, causing the star to vary as seen from Earth.*

Enthusiasts delight in the various colours of double stars. The colours are most striking when two contrasting stars are juxtaposed, as in Albireo, which marks the beak of Cygnus the swan (it is also known as Beta Cygni). Through even the smallest of telescopes Albireo reveals itself as an orange and blue–green duo, like a celestial traffic light. Swing your gaze a little further north in the summer sky and you will come to Epsilon Lyrae. Binoculars, or sharp eyes, show it as a pair of near-identical blue–white stars, but under the close scrutiny of a telescope each star divides into a close pair, from which it has earned the nickname the 'double double'.

Double stars are important since they offer astronomers one of the few accurate ways of establishing the masses of stars, which can be derived from the period of the orbit and the separation of the two members of the pair. The heavyweight champion among binaries is an otherwise unremarkable-looking star at the limit of naked-eye visibility in the constellation Monoceros, the unicorn. The Canadian astronomer John S. Plaskett, after whom it is named, was the first to recognize that this is a spectroscopic binary of exceptional mass. According to current estimates, each component of Plaskett's Star is over 55 times as massive as the Sun, and the brighter star may be as heavy as 100 solar masses.

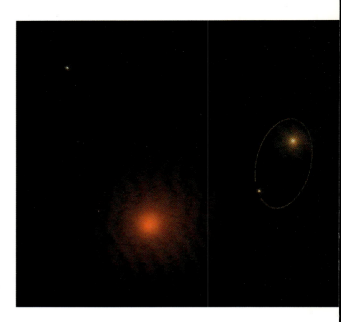

● *Charles Messier, compiler of a famous list of over 100 star clusters, nebulae, and galaxies.*

● Above left: *There are two types of double star. In some cases, one star simply lies behind the other as seen from Earth (left); this is an optical double. In other cases the two stars are actually related, with one star orbiting the other (right); this is a genuine binary.*

● Left: *The V-shaped Hyades cluster, right of centre, and, top right, the Pleiades cluster, two of the most famous open clusters in the sky. Both lie in the constellation Taurus.*

● Right: *The Double Cluster, NGC 869 and NGC 884, is a twin open cluster in the constellation Perseus, easily visible in binoculars. They are not quite identical in appearance. NGC 869 is the brighter and richer of the pair while NGC 884 contains some red giants among the predominantly blue–white stars.*

In some places, stars can be seen gathered together in clusters whose memberships range from a few dozen to many thousands. Two famous clusters lie in the constellation Taurus, the bull. One, the Hyades, is a V-shaped grouping that makes up the bull's face. The dozens of members of the Hyades are sprinkled over several degrees of sky and are best seen through binoculars. Incidentally, the red giant Aldebaran that marks the bull's glinting eye is not part of the cluster – it is a foreground object that is aligned with the Hyades by chance.

On the bull's shoulder lies another cluster that looks at first like a misty patch, rather larger than the full Moon. This is the Pleiades, or Seven Sisters, also known to astronomers as M45. Good eyesight can distinguish six or seven stars in the cluster, but it takes binoculars to reveal its true beauty – dozens of bright stars, mostly young blue giants, are suspended in space like a dew-spangled cobweb.

Many other star clusters are visible in binoculars and are marked on the star maps at the end of the book. Look, for example, for the so-called Double Cluster, an unusual twin grouping in Perseus, the constellation that represents the mythological hero who rescued Andromeda from the jaws of a sea monster. The Double Cluster, also known as NGC 869 and 884, is visible to the naked eye as a brighter knot in the Milky Way, marking the handle of the sword held aloft by Perseus. Binoculars and small telescopes will show the splash of their constituent stars, mostly blue giants but with some red giants in NGC 884.

These are all examples of *open clusters*, amorphous arrangements like a scattered handful of shiny coins. Altogether more regular are the *globular clusters*, which are spherical or elliptical in shape (see pp. 20–21). They are generally much larger than open clusters, as much as 100 light years in diameter, and contain many more stars – up to a million in the richest of them. Another difference is that globular clusters consist of old stars and are scattered in a halo around our Milky Way Galaxy, whereas open clusters contain relatively young stars and are found in the spiral arms of our Galaxy.

Although the stars in the night sky may seem as permanent and unchanging as the rocks of the Earth, many of them vary in brightness on time scales ranging from a few days to many months. The first man on record to observe a variable star was the Dutchman David Fabricius, who in 1596 watched a naked-eye star fade out of sight over a period of weeks, and did not spot it again until 1609. In the meantime it had been recorded in 1603 on the star map drawn up by the German astronomer Johann Bayer. It lay in the constellation Cetus, the mythical sea-monster to which Andromeda was offered in sacrifice. Bayer labelled this star Omicron Ceti, but it is more usually known by the name given to it in 1662 by Johannes Hevelius, who called it Mira, Latin for 'the wonderful.'

Mira is a red giant that pulsates in size, rather like a huge balloon being periodically inflated and deflated, changing in brightness as it does so. Over a period of about 11 months it goes from naked-eye brightness (usually around third magnitude, occasionally second) down to around tenth magnitude and then back up again, although the exact range of brightness, as well as the period of variation, can vary appreciably from cycle to cycle.

Stars like Mira are termed *long-period variables*, since they can take anything from two months to two years to rise and fall in brightness. Thousands of them are now known, more than for any other type of variable.

Variability is also common among red supergiants, since their extreme size makes them unstable. Their changes, however, are far less regular. Betelgeuse, for example, can go up or down in brightness by 50 per cent or more over a period of years. Antares varies too, but less noticeably.

On November 12, 1782, an 18-year-old amateur astronomer from York, John Goodricke, who had been born deaf and dumb, wrote in his observing log: "This night I looked at Beta Persei and was much surprised to find its brightness altered. I never heard of any star varying so quick in its brightness." The star he was observing is better known as Algol, the Demon Star, and it marks the head of Medusa the gorgon, carried by Perseus, who had decapitated her in one of the most famous episodes of Greek myth.

Goodricke's interest in Algol was not accidental. A York contemporary of Goodricke's, Edward Pigott, had encouraged him to keep an eye on it, for Geminiano Montanari, an Italian astronomer, had noticed its variability in 1669. Goodricke's important contribution was to establish that the variations were regular, recurring every 2 days 21 hours. Over a period of five hours Algol fades to one-third of its normal brightness, then returns to its previous level in another five hours. Goodricke could think of no other explanation "than that of supposing it to have suffered an eclipse by the interposition of a planet revolving round it."

In fact, spectroscopic observations a century later established that the eclipses were caused not by a planet but by another star, and so the Demon Star turned out to be two-headed. Algol is the type of variable known as an *eclipsing binary*. The eclipsing components of Algol are quite disparate: one is a blue–white star 25 times brighter than its companion, which is orange in colour and slightly the larger of the two. At minimum brightness about four-fifths of the brighter star is eclipsed. Half an orbit later the fainter star is eclipsed by the brighter one, but the drop in light is too slight to be noticed by the naked eye.

Eclipsing binaries such as Algol are among the most common types of variable star known. Most have periods of a few days or weeks but in one such star, Epsilon Aurigae, the eclipses occur every 27 years and last for nearly two years; the next is not due to begin until the summer of 2009.

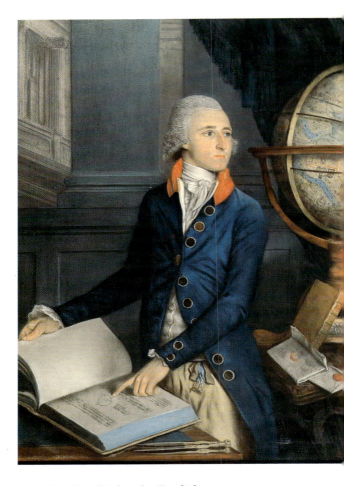

● *John Goodricke, the English amateur astronomer who began the systematic study of variable stars and discovered the variability of Delta Cephei, the prototype of the Cepheid variables.*

What was that star in the east mentioned in the Bible? Comets and novae have been suggested, but the 'star' of Bethlehem may actually have been a close approach (called a *conjunction*) between two bright planets. The possibilities have been narrowed down to a multiple conjunction between Jupiter and Saturn in 7 BC, when the two planets passed and re-passed each other three times between May and December, and an exceptionally close approach of the two brightest planets of all, Venus and Jupiter, in 2 BC. The correct identification depends on when King Herod died. Some sources place his death in 4 BC, which would point to the conjunction of 7 BC, while others believe that Herod did not die until 1 BC, which would fit the Venus–Jupiter meeting. Astronomers have calculated that Venus and Jupiter came so close that they would have appeared to merge into one brilliant 'star' on the evening of June 17, 2 BC. Furthermore, the event occurred in the constellation of Leo, the lion, which was astrologically significant to the Jews. The wise men, or *magi*, who were probably astrologers from Babylonia, would have understood the importance of such an astrological sign.

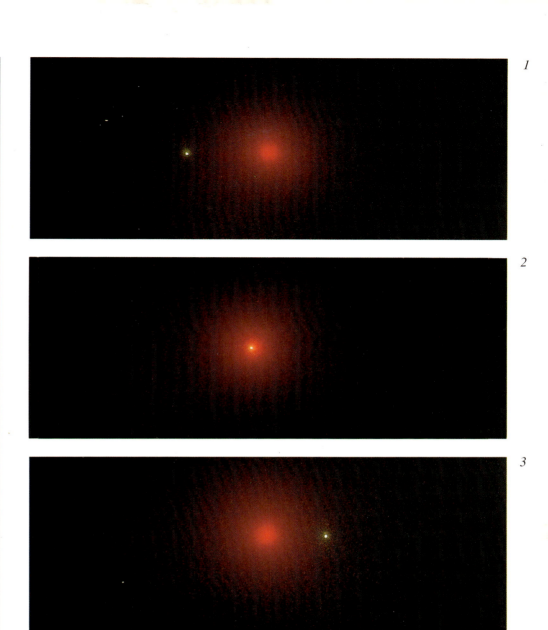

● *An eclipsing binary, of which Algol is an example, changes in apparent brightness as the stars orbit each other. When both components are visible (1 and 3) observers on Earth see the binary at maximum brightness. When the stars eclipse each other as seen from Earth (2 and 4) the total brightness drops.*

Goodricke went on to find two more variables, Beta Lyrae and Delta Cephei. Eighteen months after making these discoveries he died, apparently from pneumonia contracted while following the changes of Delta Cephei. He was 21.

Beta Lyrae and Delta Cephei, Goodricke had suggested, varied because of dark spots on their surfaces, but in this he was wrong. We now know that Beta Lyrae is an eclipsing binary, with components so close together that they are distorted into egg shapes by gravity, and gas is coiling into space from their surfaces. Delta Cephei, on the other hand, changes because it varies in size by about 10 per cent. It proved to be the most important find of all, for Goodricke had unwittingly stumbled across a rare class of stars, now known as *Cepheid variables*, that were to provide astronomers with a powerful method for finding distances in the Universe.

Cepheid variables are supergiant stars with white or yellow surfaces. Unlike red giants and supergiants, the pulsations of Cepheids repeat as regularly as clockwork, usually every few days or weeks. The variations of Delta Cephei itself take just over five days and at maximum it is twice as bright as at minimum. Another celebrated Cepheid is Polaris, the north pole star, although its variations are so slight they are barely noticeable to the naked eye. Current observations show that the pulsations of Polaris are dying out and that Shakespeare's phrase "constant as the northern star" may indeed become accurate around the year 2000.

The key fact about Cepheids, discovered in 1912 by the American astronomer Henrietta Leavitt, is that the period of their pulsations is linked directly to their brightness (i.e. their absolute magnitude) – the slowest pulsators are the brightest. Of course, the brightness that we see (i.e. the apparent magnitude) is determined by a star's distance from us, so astronomers can work out how far away a Cepheid is by comparing its calculated and observed magnitudes. For example, although Delta Cephei has the light output of 2000 Suns, it is only a modest naked-eye star because it lies 1300 light years away. Cepheid variables are the milestones that have allowed us to find the size of our Milky Way Galaxy and to deduce the distances of other galaxies.

Night after night, amateur astronomers around the world scan the dark corridors of space with binoculars, telescopes, and cameras in search of violent events. Two or three times a year they capture the outbursts of faint stars that erupt without warning to become thousands – occasionally millions – of times brighter. Such events are called *novae*, from the Latin for 'new', since astronomers of the past believed they were seeing new stars come into being.

The most spectacular nova of recent years, Nova Cygni, soared into prominence one night in August 1975, distorting the familiar cross-shape of Cygnus, the swan. During the night of discovery observers watched it double in brightness over a period of hours until it almost rivalled the constellation's brightest star, Deneb. But the glory did not last long. Over the following week it faded from naked-eye view and can now be located only in large telescopes.

Shortly after the discovery, astronomers began their usual task of trying to identify the pre-nova on old photographs. To their surprise they could not – the star had previously been so faint that it had never before been seen. Nova Cygni 1975 must therefore have risen in brightness by at least 40 million times, the greatest nova outburst on record. At its peak it shone with the radiance of a million Suns and would have been one of the brightest stars in the entire sky had it not been dimmed by intervening dust clouds.

Why should a star erupt in this way? Novae are actually close binaries consisting of a white dwarf and a red giant, or at least a star that is expanding into gianthood. Gas spills from this companion star to create a ring around the white dwarf. Gradually the gas in the ring overflows onto the surface of the white dwarf itself, eventually igniting in a nuclear explosion, throwing off gas and causing the surge in brightness.

This process can repeat itself again and again, and indeed several novae have been seen to recur, the record currently being held jointly by T Pyxidis and RS Ophiuchi, each with five observed eruptions in the past hundred years. Recurrent novae tend to have much smaller amplitudes in brightness. Perhaps Nova Cygni 1975 was so brilliant because it was erupting for the first time.

● A nova is a stellar eruption in a binary star system, one member of which is a white dwarf. Gas flows onto the white dwarf from a companion star (1 and 2), eventually igniting in a nuclear eruption (3) that throws off a shell of gas (4). This sequence of events can recur, so that the white dwarf suffers further nova outbursts.

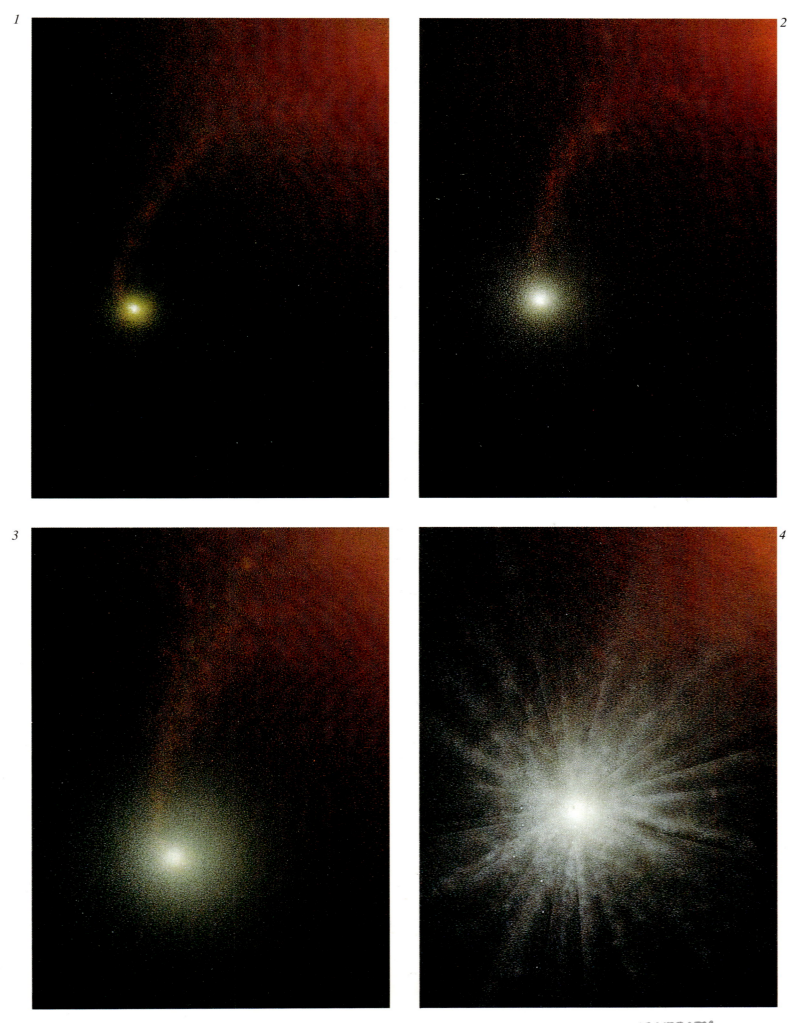

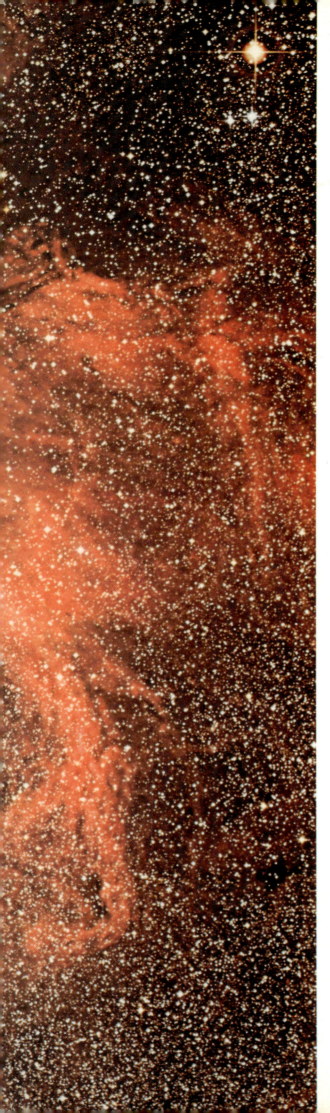

Most powerful of all stellar eruptions are the *supernovae*, which blaze hundreds of times more brightly than the brightest ordinary novae, rivalling the entire light output of a small galaxy. Supernovae are rare indeed, and one has not been seen in our Galaxy since 1604 – although in 1987 one went off in the neighbouring Large Magellanic Cloud and became a naked-eye object to southern hemisphere observers for several months.

There are two different kinds of supernova. Type I supernovae are believed to occur in binary systems and to be, in effect, the ultimate in nova eruptions – so much gas flows from the companion onto the white dwarf that the entire star detonates. Type II supernovae mark the deaths of massive single stars, as described in the next chapter.

The most likely known candidate for a supernova explosion is Eta Carinae, a star so massive that it is on the verge of instability. It has varied erratically in the past, becoming the second brightest star in the sky in the middle of the nineteenth century before fading to below naked-eye brightness, where it remains today. Stars of such extreme mass live for no more than a few million years – a very short time by stellar standards. Eta Carinae is now in the last 0.5 per cent of its lifetime, and at some time in the next 10,000 years or so the inhabitants of Earth will watch its death throes. As it tears itself apart in its final weeks of existence, Eta Carinae will dazzle the eye and cast shadows like a second Moon. Truly, no one on Earth will be unmoved by this most violent of all natural disasters.

● *Within the brightest part of this glowing nebula of hydrogen gas (above centre) lies one of the most massive stars known, Eta Carinae, 6500 light years away. Eta Carinae is an unusual variable star which may one day explode as a brilliant supernova.*

HOW ASTRONOMERS STUDY THE SKY

To get the best views of the sky, astronomers put their large telescopes on top of high mountains, above the clouds and in air that is as clear and steady as possible. All large modern telescopes are *reflectors*, which collect and focus light by means of a concave mirror; in *refractors*, a lens serves the same purpose. Mirrors are cheaper and easier to make than lenses. The largest refractor, at Yerkes Observatory in Wisconsin, has a lens 1.02 metres (40 inches) in diameter, while the largest telescope mirror, in the USSR, is 6 metres (236 inches) across.

A telescope's most vital statistic is its aperture (the width of the lens or mirror) since larger apertures collect more light, allowing fainter objects and finer detail to be seen. However, factors other than sheer size also affect the performance of a telescope, such as the quality of the optics and the steadiness of the atmosphere at the observing site, and by these criteria there are many telescopes better than the Soviet 6 metre reflector. The largest optical telescope of all, on Mauna Kea in Hawaii, has an aperture of 10 metres (32 feet), with a mirror composed of 36 hexagonal segments. In future, telescopes with single mirrors 8 metres (26 feet) in diameter will be built.

● *The 100 metre (328 feet) dish at Effelsberg near Bonn, Germany, is the world's largest fully steerable radio telescope. It was opened in 1971.*

● *The 76 metre (250 feet) radio dish at Jodrell Bank near Manchester, England, was the world's first large radio telescope, opened in 1957.*

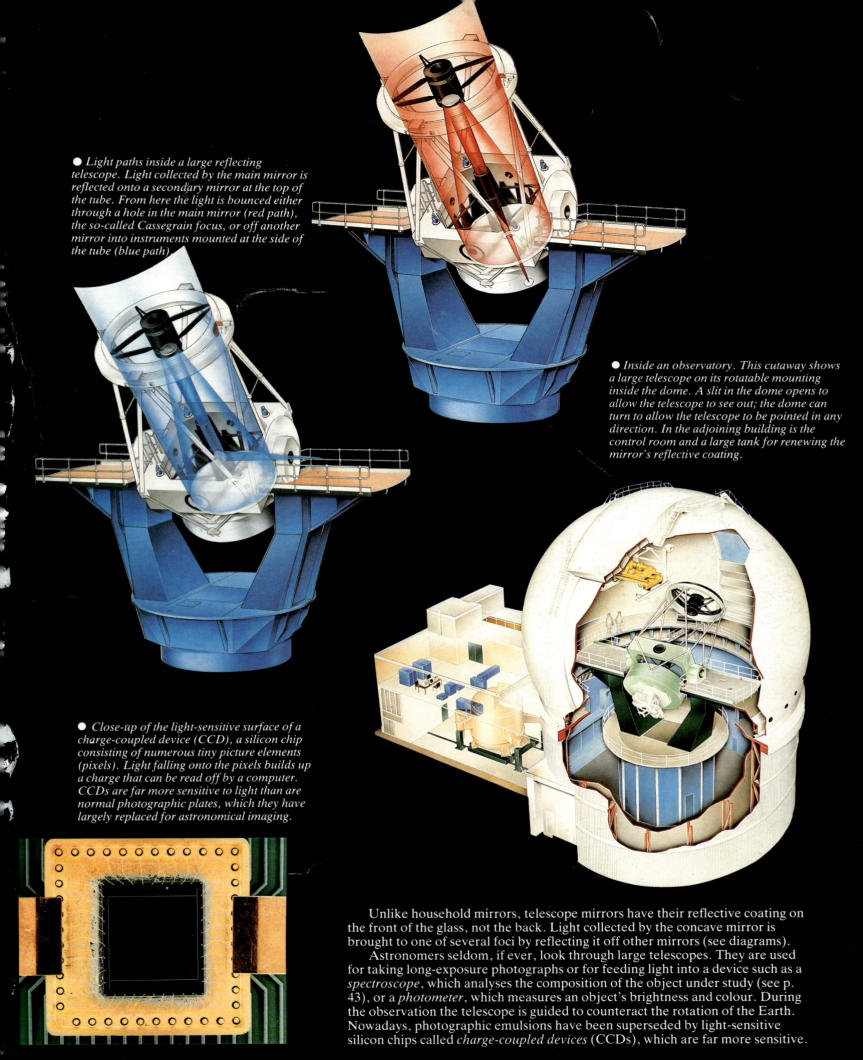

● *Light paths inside a large reflecting telescope. Light collected by the main mirror is reflected onto a secondary mirror at the top of the tube. From here the light is bounced either through a hole in the main mirror (red path), the so-called Cassegrain focus, or off another mirror into instruments mounted at the side of the tube (blue path).*

● *Inside an observatory. This cutaway shows a large telescope on its rotatable mounting inside the dome. A slit in the dome opens to allow the telescope to see out; the dome can turn to allow the telescope to be pointed in any direction. In the adjoining building is the control room and a large tank for renewing the mirror's reflective coating.*

● *Close-up of the light-sensitive surface of a charge-coupled device (CCD), a silicon chip consisting of numerous tiny picture elements (pixels). Light falling onto the pixels builds up a charge that can be read off by a computer. CCDs are far more sensitive to light than are normal photographic plates, which they have largely replaced for astronomical imaging.*

Unlike household mirrors, telescope mirrors have their reflective coating on the front of the glass, not the back. Light collected by the concave mirror is brought to one of several foci by reflecting it off other mirrors (see diagrams).

Astronomers seldom, if ever, look through large telescopes. They are used for taking long-exposure photographs or for feeding light into a device such as a *spectroscope*, which analyses the composition of the object under study (see p. 43), or a *photometer*, which measures an object's brightness and colour. During the observation the telescope is guided to counteract the rotation of the Earth. Nowadays, photographic emulsions have been superseded by light-sensitive silicon chips called *charge-coupled devices* (CCDs), which are far more sensitive.

● *Artist's impression of the Hubble Space Telescope in orbit. It was launched by the Space Shuttle in April 1990.*

The Hubble Space Telescope, launched by the Space Shuttle in 1990, has a mirror 2.4 metres (94 inches) in aperture. It can see fainter objects and finer detail than larger telescopes on Earth because it is above the atmosphere, which blurs the view of ground-based telescopes. Other satellites in space can detect wavelengths that the atmosphere filters out, such as infrared, ultraviolet, and X-rays.

Radio waves are also emitted naturally by objects in space, notably by hydrogen gas, the most abundant substance in the Universe, which radiates at a wavelength of 21 centimetres (8 inches). Since radio waves are very much longer than visible light waves, radio telescopes must have correspondingly larger apertures than optical telescopes to achieve the same definition. Most radio telescopes are shaped like dishes, which collect and focus radio waves onto a receiver. The signals are then amplified and recorded on magnetic tape.

● *The Hubble Space Telescope being released from the cargo bay of the Space Shuttle in April 1990. The golden 'wings' either side of the telescope tube are solar panels for generating electricity. To the left is the Shuttle's robot arm which was used to lift the telescope clear of the cargo bay. The telescope's aperture is covered by a protective flap in which the Earth's clouds and blue sea are reflected.*

The largest single radio dish is 305 metres (1000 feet) across at Arecibo, Puerto Rico, but it cannot be steered. The largest fully steerable dish, 100 metres (328 feet) wide, is at Effelsberg, Germany. At radio wavelengths of a few centimetres, these dishes can see the sky about as clearly as the human eye at visible wavelengths. The Arecibo dish is also used for *radar astronomy* – bouncing radio waves off objects in the Solar System. In this way information about the surface and rotation of planets, moons, and asteroids can be obtained.

One highly effective way of improving the performance of radio telescopes is to combine the output from several dishes to reproduce the view that would be achieved from one very large dish. Nowhere is this technique more spectacularly demonstrated than in the Very Large Array (VLA) in New Mexico, where twenty-seven antennae, each 25 metres (82 feet) in diameter, are arranged in a Y shape to synthesize the view of the sky that would be seen by a single dish 27 km (17 miles) across. In its definition of detail, the VLA is actually superior to the best ground-based optical telescopes.

● A view across the enormous dish of the radio telescope at Arecibo, Puerto Rico, at 305 metres (1000 feet) in diameter the world's largest single radio astronomy dish. It is suspended in a natural hollow among hills.

2 STORY OF A STAR

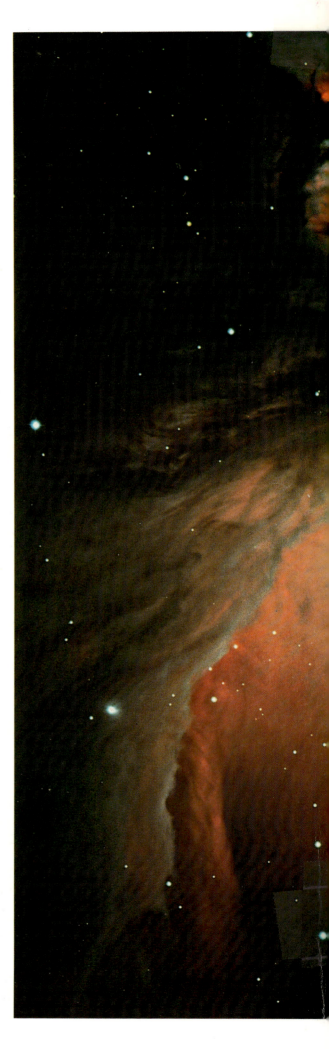

It is winter and my young Greek companion, now sensibly attired against the cold, points to a familiar constellation glittering in the south. "Orion, the hunter," she says. "Homer wrote about him in the *Odyssey*. See how he follows the Pleiades across the sky, his dogs at his heels."

Orion's stars have changed very little since the time of Homer. Betelgeuse, the red supergiant we met in the previous chapter, is embedded in the right shoulder of the hunter as he holds aloft his club in readiness to repulse the charge of Taurus, the bull. Blue–white Rigel marks Orion's left leg, and two other stars complete the outline of his body. A line of three matched stars stud Orion's belt, from which hangs his sword. Here, like a dull reflection from the smooth blade, lies the Orion Nebula, a hazy patch of gas and dust just visible to the naked eye but more distinct through binoculars.

Long-exposure photographs bring out the Orion Nebula's delicate, petal-like structure. At its centre lies a clutch of new-born stars, the Trapezium, whose youthful energy lights up the nebula that was their womb. This is the starting point in our understanding of the life-story of stars. Five billion years ago, the area of space now occupied by the Sun and planets would have looked much like the Orion Nebula.

The visible part of the Orion Nebula is only a fraction of the whole. Behind it is an extensive area of star birth, screened from human eyes by thick dust clouds but detectable by instruments sensitive to the infrared radiation (i.e. heat) given out by the warming embryo stars. The whole Orion Nebula, over 10 light years across, contains enough gas to make thousands of stars. Nebulae are composed mostly of hydrogen with a dash of helium and this, inevitably, is also the composition of stars.

A star's life-story begins when part of a nebula begins to colaesce into denser aggregations of gas. The coalescence may be precipitated by shock waves from supernova explosions of existing massive stars, thus perpetuating an endless cycle of stellar life and death. At first, the coalescing gas ball is much larger and cooler than the eventual star will be. But as it shrinks and becomes hotter, the temperature and pressure at its centre become sufficient to spark off the nuclear reactions that convert hydrogen to helium. These nuclear reactions generate the energy that makes the star shine for the rest of its life – stars are in fact immense nuclear fusion reactors.

● *Young stars lie at the centre of the Orion Nebula like new-laid eggs in a nest. The four brightest form a group called the Trapezium, visible in binoculars and small telescopes. The Orion Nebula is the most prominent of the bright clouds of gas in our Galaxy and can be seen by the naked eye on a dark winter's night. It is about 30 light years across and over 1500 light years distant.*

The ball of gas that was to become our Sun took perhaps 50 million years to contract to the nuclear-burning stage. More massive stars contract more quickly (perhaps only 10,000 years for the most massive of all), whereas low-mass stars take up to ten times longer than the Sun did to reach nuclear 'ignition'.

Nebulae like that in Orion give birth to clusters of stars. The Pleiades cluster in Taurus is an example of a relatively young star cluster, whose brightest members came into being within the past few million years. Long-exposure photographs, such as the one on the page opposite, show that they are still surrounded by traces of the nebula from which they formed. In time, clusters like this drift apart. Most likely the young Sun was once a member of a star cluster, the other members of which have long since dispersed.

Once a star settles down to stable hydrogen-burning it is said to be on the *main sequence*, a stage which the Sun reached about 4,600 million years ago. The brilliance of a star on the main sequence is directly determined by its mass, the most massive stars being the brightest. A doubling in mass produces more than a tenfold increase in brightness.

● Left: *The constellation of Orion rising on its side over rocky outcrops in Zion national park, Utah. The three stars of the belt form a nearly vertical line; the Orion Nebula is below it to the right.*

● Below: *A star is born when a cloud of gas starts to collapse, perhaps triggered by the shock waves from a nearby supernova explosion (top left). As the protostar collapses, it begins to glow a dull red; it is surrounded by a disk of matter that may later form into a planetary system.*

● Right: *The Pleiades star cluster, popularly known as the Seven Sisters, a group of young, blue giants still surrounded by the gauzy remains of the nebula from which they were born.*

As we saw in Chapter One, star masses range from about a hundred Suns to just under a tenth of a Sun. The most massive stars on the main sequence are also the hottest, and hence the bluest; they are millions of times brighter than the Sun, whereas the cool red dwarfs at the lower end of the main sequence are over 10,000 times fainter. Low-mass stars are in fact the most abundant, but they are the least conspicuous because they are so faint. Although the most massive stars on the main sequence are much larger than the Sun, they are not yet giants or supergiants: stars do not reach that stage until they grow old, as we shall see.

NEBULAE – CLOUDS IN SPACE

Clouds of gas and dust are dotted throughout the spiral arms of our Galaxy. Some, such as the famous Orion Nebula, glow from the ultraviolet light emitted by hot, young stars within them, while others are dark. Glowing nebulae usually appear pink or red on photographs, that being the colour of light emitted by their main constituent, hydrogen gas. One of the largest nebulae known is the Tarantula Nebula, a spider-shaped mass of gas that lies 160,000 light years away in the Large Magellanic Cloud, a close neighbour of our Galaxy. If the Tarantula were as close to us as the Orion Nebula is, it would fill the entire constellation of Orion and cast shadows at night.

Dusty nebulae are lit up by a different process: they simply reflect light from nearby stars, scattering it to produce a sky-blue colour. The contrast in colour between adjacent patches of glowing gas and reflecting dust is well shown in the Trifid Nebula.

Nebulae that remain dark can be seen only where they are silhouetted against a brighter background, either a Milky Way star field or another nebula. The Coalsack, in the southern constellation of Crux, the southern cross, gives the impression of being a dark hole in the Milky Way. Most remarkable of all is the Horsehead Nebula, a finger of dust on a backdrop of bright red nebulosity near Orion's belt (but not part of the better-known Orion Nebula). The Horsehead is faint, and shows up well only on long-exposure photographs.

Planetary nebulae (see pp. 50–51) are small, usually rounded clouds formed by the deaths of stars like the Sun. Some other nebulae are formed by the supernova explosions of massive stars, the Crab Nebula (see pp. 52–53) and Veil Nebula being good examples.

● Above left: *The flower-like Rosette Nebula, a loop of gas in the constellation Monoceros encircling a cluster of stars. The Rosette is estimated to be about 100 light years in diameter.*

● Above: *The Eagle Nebula in Serpens is a mass of star spawn surrounding the star cluster M16. Egg-like knots of gas in the nebula will condense into stars.*

● Above right: *The Trifid Nebula, M20, in Sagittarius, takes its name from the three lanes of dark dust that trisect its pink gas cloud. Another part of the nebula is blue because of starlight scattered off dust particles.*

● Right: *The dark Horsehead Nebula, a celestial chess-piece, is a cloud of dark dust silhouetted against a background of glowing red hydrogen gas.*

● Left: *The Lagoon Nebula, M8, in Sagittarius, another example of a combined gas cloud and star cluster, is bright enough to be easily visible in binoculars.*

Our Sun is a typical star, and its statistics can be summed up easily. Mass: 330,000 times that of the Earth. Diameter: about 1.4 million km (900,000 miles), or 110 times the Earth's diameter. Average density: 1.4 times the density of water (much higher at the centre, much lower at the surface). Volume: 1.3 million times that of Earth. Surface temperature: 5500°C. Distance from Earth: 150 million km (93 million miles); light takes 8.3 minutes to travel this distance.

Yet these bald figures conceal the fact that the Sun is a seething ball of gas in constant turmoil, racked by storms and blemished by dark spots. It is of particular interest to astronomers since it is the only star we can see in any detail, and of vital importance to everyone on Earth since it provides the light and heat to keep us alive. If anything unusual were about to happen to the Sun – for example, a change in its energy output that would affect the Earth's climate – we would want some warning. Astronomers poring over dusty historical records have found worrying evidence that the Sun may be slightly variable, a finding backed up by modern satellite observations, as we shall see.

Unlike the stars of night, which appear as nothing more than points of light through the largest ground-based telescopes, we can see features on the face of the Sun with the simplest optical equipment. **Never look directly at the Sun through binoculars or telescope for you will blind yourself** – anyone who has used a magnifying glass to focus sunlight onto paper will appreciate the burning powers of the Sun's rays. The only safe way to observe the Sun is to project its image onto a white surface.

Most immediately noticeable are dark sunspots, transient features that come and go with lifetimes of a few days or weeks, although some can last for months. They develop from day to day, and the largest can be seen with the naked eye when the Sun is low and dimmed by haze. A typical sunspot could swallow several Earths.

Sunspots were first recorded long before the invention of the telescope, witness this Chinese record from AD 188: "The Sun was orange in colour. Within it there was a black vapour like a flying magpie. After several months it dispersed." Sunspots were also recorded by astronomers in Arabia, as in this account from AD 840: "In the year 225 [by the Muslim calendar] during the Caliphate of al-Mu'tasim there appeared a black spot close to the middle of the Sun . . . this spot persisted on the Sun for 91 days."

It took a pharmacist from Dessau, Germany, to show that there was a pattern to the appearance of sunspots. Heinrich Schwabe patiently observed the Sun's face with a small telescope every clear day for 17 years, and by 1843 he was convinced that the number of spots rose and fell in a cycle lasting about 10 years. Schwabe's surprise finding marked the beginning of solar physics. It inspired Rudolf Wolf at Zurich Observatory to begin a programme of solar monitoring that continues to this day.

Wolf established that the average length of the sunspot cycle is nearer 11 years, although the exact length varies considerably from cycle to cycle, as does the number of sunspots. He counted the number of spots and spot groups visible on the face of the Sun to produce an index of solar activity known as the Wolf sunspot number, still used today. At solar minimum, the Sun's disk may have no spots visible on it for days on end, whereas around maximum dozens of them give it the appearance of a speckled egg. At the highest recorded maximum, in 1957, the Wolf sunspot number reached 190, whereas at the low maxima of 1804 and 1816 it peaked below 50.

Another man inspired by the work of Schwabe was the English amateur astronomer Richard Carrington, who in 1853 began to measure the rotation of the Sun by watching the motion of sunspots. As the Sun rotates, spots appear around one edge (called the *limb*), are carried across the Sun's face, and disappear at the opposite limb. Carrington

● *To observe the Sun and its spots safely, project its image with telescope or binoculars onto a white surface. To look directly at the Sun through any form of optical instrument is to risk being blinded.*

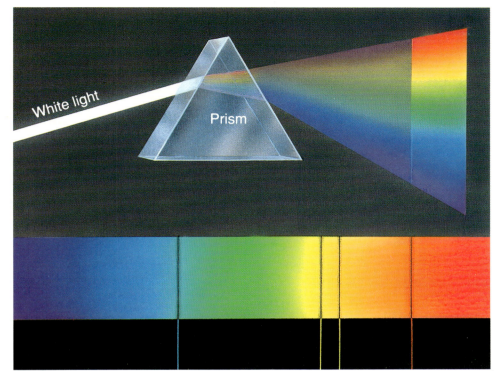

White light

Prism

● Above: *Joseph von Fraunhofer, the German scientist who discovered the dark lines in the Sun's spectrum caused by the presence of various gases.*

● Above right: *The formation of spectra. Light passed through a prism is spread out into a rainbow-like band of colours. The spectra of stars are crossed by dark lines, called Fraunhofer lines, which reveal the composition of the stars. Nebulae have bright emission lines in their spectra.*

MEASURING THE COMPOSITION OF THE STARS

It is possible to analyse the composition of stars and galaxies without ever leaving Earth — by studying their light. When light from any bright object is passed through a prism, the different wavelengths are spread out into the rainbow band of colours known as the *spectrum*, longest wavelengths at the red end and shortest wavelengths at the blue end. A *spectroscope*, the instrument used by scientists to analyse light, works on this principle. Much astronomical observation consists of taking spectra. The Sun's spectrum (and those of stars) is crossed by a series of dark, threadlike lines caused by the absorption of various wavelengths by cooler gas in the Sun's outer layers. These are called *Fraunhofer lines*, after the German physicist Joseph Fraunhofer who first studied them in 1814. Each chemical element has its own distinctive pattern of lines, from which we can deduce its presence in the object under study. In this way the Sun has been found to be made up of just over 90 per cent hydrogen (by number of atoms) plus nearly 10 per cent helium, other elements accounting for less than 1 per cent of the total.

Nebulae, too, have had their composition analysed spectroscopically. As they are much less dense than stars, their spectra consist of isolated bright lines. Astronomers can deduce many things other than chemical composition from spectra, including the presence of magnetic fields, and also how objects are moving; for instance an apparently solitary star may be unmasked as a close binary — a spectroscopic binary — through the effect of the orbital motion on its spectrum.

found that the Sun rotates in about 25 days near the equator, but that halfway to the poles this has slowed to 27½ days. Sunspots are seldom seen further from the equator than this, but other ways have been found to measure the rotation of the Sun at higher latitudes; it slows to about 36 days close to the poles. (Such a *differential rotation* arises because the Sun is a gaseous body, not solid.) Only the largest spots persist for more than one solar rotation, as happened with the ancient naked-eye sightings quoted on page 42. The longest-lived sunspot group in recent times lasted seven months, from June to December 1943.

Sunspots are actually areas of cooler gas that appear dark by contrast with the hotter surface, the *photosphere*, surrounding them. Spots come in a variety of shapes, from simple rounded ones to complex mother-and-children groupings. A sunspot's dark core, the *umbra*, has a temperature of around 4000°C, similar to that of an orange giant star such as Arcturus. Surrounding this is the lighter *penumbra*, usually much larger and about 1000°C warmer than the umbra. Large groups can extend like island chains for 100,000 km (60,000 miles) across the Sun's surface. A monster sunspot group that appeared in April 1947 measured 300,000 km (190,000 miles) in length, enough to stretch most of the way from the Earth to the Moon, and could have contained a hundred Earths.

Sunspots occur where the magnetic field of the Sun bursts through the surface like strands from a tangled ball of wool. Although details remain obscure, it seems that the differential rotation of the Sun twists the solar magnetic field until it runs almost east–west beneath the photosphere. The intense local magnetic fields, thousands of times stronger than the Earth's, hinder the convection of heat where they penetrate the photosphere, creating cooler patches. As the magnetic field dissipates the spots fade away.

Large sunspots often occur in pairs, oriented east–west, which act like the north and south poles of a horseshoe magnet, one pole where the magnetic field emerges from the Sun and the other pole where it re-enters. Throughout each 11 year cycle, the leading spot of a pair in a given hemisphere always has the same magnetic polarity, opposite to the polarity of the leaders in the other hemisphere. In 1912 the American astronomer George Ellery Hale at Mount Wilson Observatory, California, was surprised to find that the magnetic polarities of the leading and following spots in each hemisphere had switched over with the start of a new sunspot cycle. Continued observations showed that the polarities switched back again at the start of the subsequent cycle. Hale had established that the true length of the solar cycle is not 11 years after all, but 22 years.

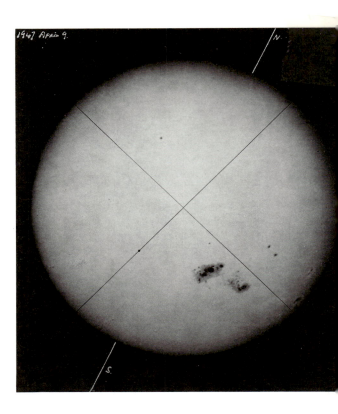

● *The giant sunspot group of April 1947, one of the largest ever seen, stretched for 300,000 km (200,000 miles), almost as far as from the Earth to the Moon. The intersecting lines on the photograph are cross-wires in the telescope eyepiece.*

● Left: *George Ellery Hale discovered that sunspots and other activity on the Sun were linked with strong magnetic fields. Hale (third from left) is pictured here with a group of other astronomers on Mount Wilson, California, where he set up solar telescopes and the Mount Wilson 2.5 metre (100 inch) reflector.*

Closely associated with sunspots are localized eruptions on the Sun's surface called *flares*. Richard Carrington caught sight of a flare by chance while routinely observing the Sun on the morning of September 1, 1859. He noted two patches of bright, white light that quickly rose to a peak of intensity and died away again after little more than a minute. The following night there were strong disturbances of the Earth's magnetic field, and people in many parts of the world reported colourful glows in the sky called *aurorae*.

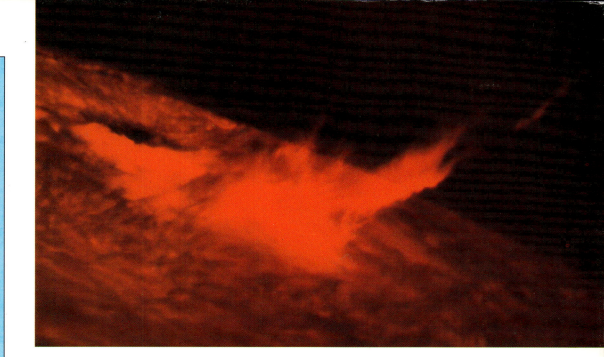

THE STELLAR POWERHOUSE

Stars do not burn in the way that things burn on Earth. Rather, their energy comes from processes involving the very centres, the nuclei, of atoms — that is to say, nuclear reactions. It was not until the 1930s, with the development of nuclear physics, that astrophysicists were finally able to explain what makes the Sun and the stars shine.

Inside stars, four protons (atomic particles that are the nuclei of hydrogen atoms) are fused together to make the nucleus of one helium atom. A helium nucleus, though, has fractionally less mass than four hydrogen nuclei and the difference in mass is released in the form of energy. The mass difference is less than 1 per cent but the energy it produces is enormous — this is the same process that powers a hydrogen bomb.

At the core of our Sun, where the temperature is 15 million °C and the pressure is 300 billion atmospheres, 600 million tonnes of hydrogen is converted to helium every second, and 4 million tonnes turned into energy. The Sun started life with enough hydrogen to keep it burning for around 10,000 million years in all, even at this prodigious rate. It is currently about halfway through its life.

Nuclear fusion is not to be confused with fission, which involves the splitting of heavy atoms and is the process currently used in nuclear power stations on Earth. Scientists are working to harness nuclear fusion, the energy source of the stars, in a controlled way to provide clean energy for the future, replacing the current generation of fission reactors.

● Above right: *A solar flare ejects hot gas at high speed from an explosion on the Sun's surface.*

● Right: *The green and red glows of an aurora seen over Fairbanks, Alaska, caused by atomic particles from the Sun bombarding the Earth's upper atmosphere.*

This was the first evidence that solar activity affects the Earth, but only in recent years have we come to understand how it happens. Flares are caused by a sudden release of magnetic energy above a sunspot. They fire out a high-speed salvo of atomic particles, mostly electrons and protons. When these particles reach the Earth they are funnelled by the Earth's magnetic field towards the poles where they rain down on the upper atmosphere, making it glow green and red like a multi-coloured fluorescent tube at heights above 100 km (60 miles). Aurorae, popularly known as the northern (or southern) lights, can take on amazing shapes such as arches and folded curtains that gently ripple as though being blown by a breeze.

In response to disturbances of the Earth's magnetic field caused by the arrival of the atomic particles, electric currents are induced in conductors at the Earth's surface such as power cables and oil pipelines. This can lead to interruptions in power supplies, disturbance of telephone networks, and corrosion of pipelines. These effects are most commonly felt at high latitudes, as in Canada, Alaska, and Scandinavia. The greatest auroral display of recent years took place on the night of March 13, 1989, following a major flare on the Sun. Colourful aurorae were seen throughout Europe and the USA, and as far south as the Caribbean. In Canada, surges on power lines interrupted the electricity supplies throughout the province of Quebec.

● Left: *The Sun's corona, a faint halo of gas boiled off its surface, becomes visible at a total solar eclipse. Here it is seen at the total eclipse of February 16, 1980, from Hyderabad, India.*

● Below left: *A prominence – a stream of hot gas arching into space from the surface of the Sun, photographed in the red light of hydrogen.*

● Below: Frost Fair on the Thames *by Abraham Hondius, painted in 1677 in the middle of the Little Ice Age when temperatures in northern Europe were exceptionally cold.*

Other forms of activity on the Sun rise and fall in step with the number of sunspots. Most notable are *prominences*, huge clouds of gas that arch above the Sun's surface along lines of magnetic force. Prominences can be seen at the Sun's limb at a total eclipse, or with special instruments that block off the light from the Sun's disk. When seen silhouetted against the photosphere they appear as dark ribbons, called *filaments*. Some prominences persist little changed for weeks while others, associated with flares, blow away from the Sun at high speed.

Earlier, we referred to the possibility that the Sun might be slightly variable. Between 1645 and 1715 hardly any sunspots or aurorae were seen, a period that is called the Maunder minimum after E. Walter Maunder, a British astronomer who pointed it out at the end of the nineteenth century. Coincidentally, during the Maunder minimum the Earth was unusually cold, a period now termed the Little Ice Age, when glaciers expanded, crops failed, and the rivers of Europe habitually froze in winter. ('Frost fairs' were held regularly on the frozen Thames, notably in 1684 when the river stayed frozen for two months.) Was this cooling of the Earth associated with the lack of sunspots? If so, how?

Long-term prediction of solar activity is a major goal for astronomers, and they are being helped by satellites that have provided direct evidence that the Sun's output varies in step with the sunspot cycle. From 1980 to 1986, instruments aboard the Solar Maximum Mission and Nimbus 7 satellites recorded a gradual decline in the Sun's luminosity as sunspot activity fell from maximum to minimum. When the number of sunspots increased again after 1986, the Sun's luminosity rose accordingly.

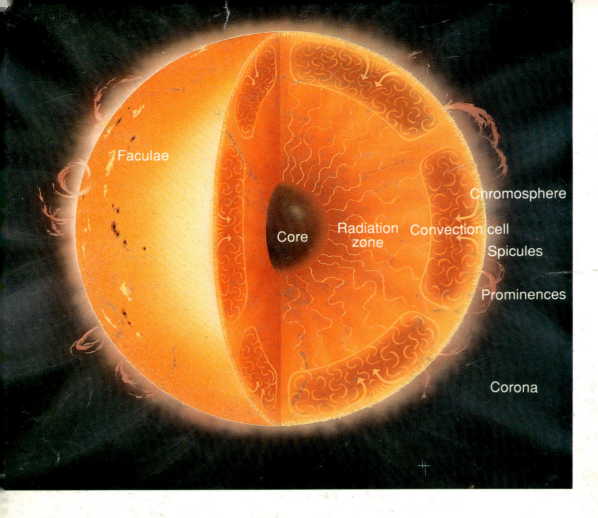

● Energy generated at the Sun's core
first moves outwards by radiation
(represented by the zigzagging lines)
and then by convection cells of gas. On
the Sun's surface are dark spots, bright
faculae, and looping prominences.
Beyond these are the pinkish
chromosphere and the pearly corona.

Why should the Sun be brightest when sunspots are most numerous? The answer lies with bright patches on the photosphere called *faculae*, which precede the formation of sunspots and remain for some weeks after the spots have gone. Their brightness more than compensates for the dimming caused by the increased number of sunspots.

Between the sunspot maximum of 1980 and the minimum of 1986, the Sun's luminosity decreased by about 0.1 per cent, small but not negligible. Calculations predict that such a change would cause a drop of less than 0.1°C in the average temperature of the Earth, only a fraction of the warming caused by the increased carbon dioxide content of the atmosphere. However, a reduction of only about 0.5°C in the Earth's average temperature would be enough to precipitate another Little Ice Age and that, solar physicists believe, could happen if the Sun's luminosity were to drop by 0.5 per cent over several sunspot cycles. From what we know of the Sun's past behaviour during the Maunder minimum – no sunspots for 70 years, and presumably no faculae either – such a dip in luminosity may well be possible. All the more reason to keep a careful watch on the Sun.

Whatever the short-term prospects for the Sun, the long-term forecast is clear. Imperceptibly at first, it will brighten and swell over the next few billion years until it is twice as bright and perhaps 50 per cent bigger than at present. Long before then, the Earth will have become uninhabitably hot; the polar caps will have melted, and the most fertile land will have become desert. Fortunately, this is not something we need to worry about for a few hundred million years yet.

Eventually, about 5000 million years from now, the Sun will run out of hydrogen in its core. This will set off an accelerating series of changes that will transform the star, and ultimately lead to its death. First, the nuclear reactions move outwards into the hitherto untouched hydrogen surrounding the burnt-out core, which shrinks. The Sun is about to leave the main sequence, never to return.

STRUCTURE OF THE SUN

Energy is generated in the Sun's core, about 400,000 km (250,000 miles) across. The energy radiates outwards from atom to atom to within about 100,000 km (60,000 miles) of the surface, completing its journey to the surface by convection in huge rising currents of gas. Eventually the energy from the core appears as heat and light at the visible surface, the *photosphere*, a sea of boiling gas about 300 km (200 miles) deep. Seen in close-up, rising and falling currents bubbling to the surface produce a mottled effect known as *granulation*, each granule being about 1000 km (600 miles) across. Dark *sunspots*, actually cooler areas, blemish the photosphere; typically about 20,000 km (12,000 miles) across, they can occur in groups and chains that stretch much further.

Above the photosphere is a more tenuous layer, the *chromosphere*, through which tongues of gas called *spicules* dart to a height of about 10,000 km (6000 miles), giving the chromosphere the appearance of a flaming forest. Higher still are *prominences*, clouds of denser gas that follow lines of magnetic force — although 'denser' here is purely relative, since they have only a fraction of the density of the Earth's atmosphere. Beyond the chromosphere is the *corona*, a faint and exceedingly rarefied halo of gas that has boiled off from the Sun. Gases from the corona stream outwards, forming the *solar wind* of atomic particles that blows through the Solar System. In a sense, the Earth and other planets orbit within the outermost reaches of the corona.

● *Billions of years into the future the red giant Sun will loom large in the sky, its fierce heat melting the lifeless Earth into a ball of lava. Eventually the Sun may swell enough to swallow the Earth.*

These internal changes release more energy than before, pushing the overlying layers of gas outwards. As the Sun expands, its surface becomes cooler and redder. It has turned into a red giant, dozens of times larger than it is now, filling Mercury's orbit and shining hundreds of times more brightly than today. It will resemble the red giant Arcturus. When the Sun swells into a red giant the oceans of Earth will boil away, leaving our planet a parched cinder. The rocks at the Earth's surface will glow like the walls of a blast-furnace.

If we could dive into the red giant Sun, we would find that its outer layers are far more tenuous than our own atmosphere. By the standards of an Earth laboratory, the outer layers of a red giant are a good vacuum. Even at the speed of our fastest space probes it would take several weeks to plunge the 50 million km (30 million miles) through the star's outer atmosphere.

Eventually we would come within sight of the star's white-hot core, perhaps 30,000 km (20,000 miles) across, somewhat smaller than the planets Uranus and Neptune. By contrast with the near-vacuum of the red giant's outer layers, the helium core has a density of several tonnes per cubic centimetre. Around it is a thin shell of burning hydrogen a few thousand kilometres deep.

This core, tiny by comparison with the overall dimensions of the red giant, contains fully half the mass of the entire Sun. If a red giant were shrunk to fit into your living-room, its core would be the size of this letter o.

After some millions of years in this state, the contracting helium core in the red giant Sun becomes hot enough – about 100 million °C – to initiate the fusion of helium into carbon and oxygen. Events now proceed as they did in the earlier part of the Sun's life, only much more rapidly. Carbon 'ash' collects at the core; the core begins to contract and the helium burning migrates outwards, still with a hydrogen-burning layer above it. The red giant Sun will now swell even more, this time engulfing what remains of the Earth.

The end is now not far away. So distended does the Sun become that over the ensuing 100,000 years or so its outer layers will drift away into space, forming a huge smoke-ring that envelops what was once the Solar System. Remarkably enough, we can see the ejected gas shells of about a thousand stars that have reached this stage in their evolution. They are known as *planetary nebulae*, because they look like the disk of a planet when seen through a small telescope. The name is entirely misleading, for they have nothing to do with planets at all, but we are stuck with it.

At the centre of the expanding gas shell is a faint star, the core of the former red giant that is now exposed to view. Over millions upon millions of years this core shrinks and fades to become a *white dwarf*, a tiny but hot and very dense star. About one per year is formed in the Milky Way. When the Sun becomes a white dwarf it will have about half its current mass squeezed into a ball the size of the Earth, the rest of the mass having been lost into space during the planetary nebula stage.

The first white dwarf was discovered by accident in 1862 by an American optician, Alvan G. Clark, while he was testing a new telescope. He turned it towards the star Sirius and saw, to his surprise, that Sirius had a faint companion, never previously seen. However, the discovery was not entirely unanticipated. Some twenty years earlier, the German astronomer Friedrich Wilhelm Bessel had analysed the motions of Sirius and Procyon, the brightest stars in the constellations Canis Major and Canis Minor. He found that both stars followed wavering paths through space, which he attributed to the presence of an invisible companion to each. Clark's telescopic discovery showed that the companion of Sirius, at least, was not invisible – just difficult to spot

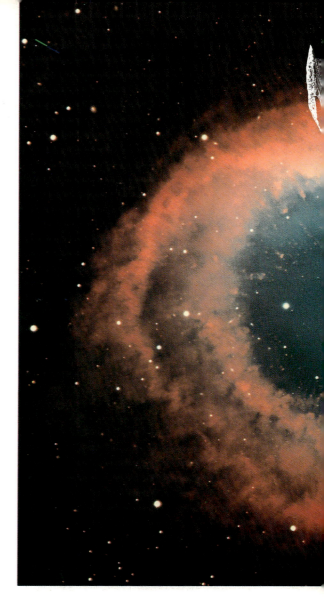

● Above: *The Helix Nebula in Aquarius, the largest of the planetary nebulae in apparent size, is about 450 light years distant.*

● Above right: *M57, the Ring Nebula in Lyra, is over 2000 light years away. It is a good object for moderate-sized telescopes.*

● Right: *About 700 light years away in Vulpecula lies the Dumbbell Nebula, M27. It is easily visible in binoculars.*

because it was so faint and so close to the dazzling Sirius itself. In 1896 the companion to Procyon was also seen, by John M. Schaeberle at Lick Observatory, California.

Astronomers realized that these companions were unusually faint for their calculated masses, but the full story did not become clear until 1915 when the American astronomer Walter S. Adams measured the surface temperature of Sirius B, as the companion is known, and found it to be unexpectedly hot. Here was a contradiction – a star hotter than the Sun, with a mass similar to the Sun's, but very much fainter. The only possible answer was that the star must be unprecedentedly small, and hence extraordinarily dense. We now know that Sirius B is twice the size of the Earth and 100,000 times the density of water, far denser than any material known on Earth. A teaspoonful of matter from Sirius B would weigh several tonnes.

At the core of a white dwarf, atomic particles are squashed together as closely as is physically possible, leading to the incredible densities. No energy is being generated inside white dwarfs, since nuclear reactions have ceased. Over billions of years the stars radiate away their remaining heat, finally fading into oblivion.

Stars up to five or six times the Sun's mass go through the same evolution as the Sun, although more quickly, for the lifetime of a star – like so much else – depends on its mass. Surprisingly, the most massive stars have the shortest lifetimes, since their energetic lifestyles require a high fuel consumption. A star twice the mass of the Sun, such as Sirius, has one-tenth the Sun's lifetime, whereas the most massive stars of all live for only a few million years, a fleeting existence by astronomical (and geological) standards.

Massive stars not only live fast and die young, they go out with a bang. They follow a different path to their destiny from less massive stars, starting when they evolve into supergiants, which as we have seen are distinguished from ordinary giants by their luminosity rather than their size. Supergiants are so hot at their centres that fusion reactions progress far beyond the helium-to-carbon stage at which Sun-like stars stop. A runaway sequence of reactions ensues, building up progressively heavier elements, which ends with the fusion of silicon into iron. (It should be noted that the temperatures at the star's centre are so great that the iron is not solid as we know it on Earth.)

In the energy stakes, iron is a worthless commodity. Fusion of iron does not release energy, it absorbs it – so a star with an iron core is like a house without foundations. With its central energy source gone, the star suddenly and catastrophically collapses in upon itself, triggering an explosion like an immense nuclear bomb. The star has become a supernova, one of the most violent events in nature. For a few weeks the exploding star shines as brightly as 100 million Suns, rivalling the light output of an entire galaxy.

All the chemical elements of nature are fashioned in the nuclear furnace of a supernova, and are scattered into space where they mix with existing nebulae, later to be collected up into new stars, and perhaps planets with life. The very atoms of our bodies were produced in the supernova explosions of stars that lived and died before our Sun was born.

Chinese astronomers recorded a brilliant supernova in AD 1054, which they termed a 'guest star'. According to contemporary Chinese accounts, the guest star was bright enough to be visible in daylight for 23 days, and could be seen by the naked eye for over 18 months. At the site of the Chinese guest star, between the horns of Taurus, the bull, lies one of the most celebrated objects in the heavens – a patch of gas known to modern astronomers as M1, the first entry in the list of nebular objects compiled by the French comet hunter Charles Messier. It is popularly

● Below: *The Crab Nebula, the shattered remains of a star that was seen to explode in AD 1054. The pale blue light from the central part of the nebula is produced by electrons spiralling in the nebula's intense magnetic field. Superimposed on this are reddish filaments of hydrogen gas flung off in the supernova explosion. At the Nebula's centre is a close pair of stars. The lower one is a pulsar, the remains of the star that exploded; its energy keeps the nebula glowing. The other star is an unrelated foreground object.*

● Above: *The pulsar in the Crab Nebula flashes on and off 30 times a second, as seen in this sequence of high-speed photographs.*

● Below right: *Splashed across the southern constellation of Vela are filaments of gas from a supernova explosion about 10,000 years ago, appearing like cirrus clouds lit up by the setting Sun. The explosion created a pulsar that flashes 12 times a second.*

known as the Crab Nebula, because the nineteenth-century Irish astronomer Lord Rosse drew it with crab-like claws, although the resemblance is difficult to see.

The Crab Nebula is the wreckage of the 1054 supernova, still flying apart at speeds of up to 5 million kph (3 million mph). Its current diameter is about 10 light years, and it is estimated to lie 6300 light years away – which means that the light from the explosion took 6300 years to reach the eyes of the Chinese astronomers.

Supernovae are rare in our Galaxy, although we can see them through telescopes going off from time to time in other galaxies, so when in 1987 the first naked-eye supernova for nearly 400 years shone forth, astronomers dropped whatever they were doing to study it. Ian Shelton, a Canadian astronomer working at Las Campanas Observatory in Chile, spotted it in the early hours of February 24 on a photograph he had taken of the Large Magellanic Cloud, a satellite galaxy of our Milky Way. There, next to the Tarantula Nebula, was a bright new star, now known as Supernova 1987A. It remained visible to the naked eye until the end of that year, delighting professional astronomers and public alike. Telescopes throughout the southern hemisphere were turned towards it, as well as satellites in space. Supernova 1987A was the most important event in many astronomers' lives.

● *In February 1987 the brightest stellar explosion for nearly 400 years, Supernova 1987A, went off in the Large Magellanic Cloud, a neighbouring galaxy of the Milky Way. It was visible to the naked eye for over six months. Here, photographed shortly after the outburst, it glows orange at lower right near the Tarantula Nebula, so named because of its spidery appearance.*

Old photographs of the area were scanned to identify the star that exploded. It was a supergiant of about 20 solar masses, as expected, but there was a surprise – it had not been red at all, but blue. Was the theory of supernovae wrong? Probably not. Most likely the star evolved into a red supergiant a few thousand years ago but then shed its outer layers, exposing its hotter, bluer interior.

A supernova explosion flings off the outer layers of a star, leaving behind the core, but in a vastly altered state. Under the immense hammer blow of a supernova, the electrons and protons at the supergiant's core are melded to form neutrons. The resulting object is known, logically enough, as a *neutron star*, even smaller and denser than a white dwarf.

The existence of stars more exotic than white dwarfs was predicted during the 1930s. The Indian astrophysicist Subrahmanyan Chandrasekhar showed that there was an upper limit to the mass of white dwarfs above which their own gravity would be so strong that they would inevitably collapse into something smaller and denser. This critical mass, 1.4 Suns, is now known as the Chandrasekhar limit.

There are plenty of stars heavier than this, with little hope of them shedding enough mass as either planetary nebulae or supernovae to duck below it. We now know that they end up as neutron stars, but when J. Robert Oppenheimer, better known as the father of the atomic bomb, calculated the properties of neutron stars in 1939, most astronomers dismissed the results as fantasy. Such a star would have more mass than the Sun compressed into a ball perhaps 20 km (12 miles) across. The resulting density would be so great that a teaspoonful of the material would weigh perhaps a billion tonnes. Neutron stars would be so small and faint that there seemed no chance of detecting them, even if they existed.

That was where the matter remained for almost thirty years. What changed things was a recurring blip on a pen-chart linked to a radio telescope at Cambridge, England. A student, Jocelyn Bell, noticed the signal and brought it to the attention of her supervisor, Antony Hewish. More detailed study confirmed that the signal consisted of radio pulses that arrived, every 1.3 seconds, with the regularity of a highly accurate clock. Soon, several other rapidly pulsating sources were discovered. They were christened *pulsars*.

● *Antony Hewish and Jocelyn Bell, discoverers of the flashing stars known as pulsars.*

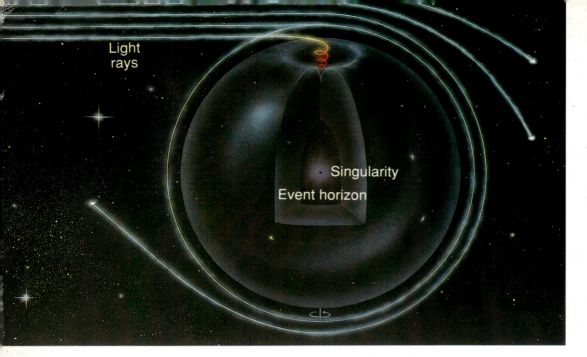

Light rays

· Singularity

Event horizon

To flash on and off as quickly as a pulsar, the object had to be very small indeed by stellar standards. It was not long before astronomers realized that they had, at last, stumbled upon neutron stars. The link was confirmed beyond doubt in 1968 when a pulsar was detected in the Crab Nebula, and the following year this star was seen flashing optically, exactly in step with the radio pulses – it is the lower of the two stars at the centre of the nebula. Unfortunately the flashes, at 30 a second, are too fast for the eye to detect. Several hundred pulsars have now been discovered, most of them at radio or X-ray wavelengths. The Crab pulsar remains the brightest, and one of the very few to have been detected optically.

Pulsars may be thought of as celestial lighthouses, emitting beams of energy that produce a flash every time they turn. The Crab pulsar, for example, must be spinning 30 times a second, and the fastest pulsars are spinning hundreds of times a second. Only an object as small and dense as a neutron star could spin so quickly without disintegrating.

The discovery of neutron stars revitalized interest in an even more bizarre prediction made by some theorists that until then had seemed exclusively the province of science fiction writers – *black holes*. According to the theory, if a neutron star has a mass of more than about three Suns (the exact limit is not precisely established) then nothing can stop it from shrinking under the inward pull of its own gravity until it vanishes from sight into a black hole.

At the centre of the black hole, the irresistible force of gravity compresses the former star to infinitely high density at an infinitely small point, called a *singularity*. The star has crushed itself not just out of sight but literally out of existence. Around this singularity is an intense gravitational field that prevents the escape of anything that should trespass into it. Even the star's own light cannot escape from this exclusion zone, known as the *Schwarzschild radius*. Karl Schwarzschild was a German astrophysicist who calculated that black holes should exist. This was while he was studying relativity theory in 1916, long before astronomers took such theoretical oddities seriously. For a black hole with the mass of the Sun the Schwarzschild radius is about 3 km (2 miles); it is bigger for greater masses.

Since many supernovae must leave cores of more than three solar masses, black holes should be plentiful throughout the Galaxy – but how do we find them? Fortunately, some black holes occur in binary systems and so give themselves away by the effects they have on their visible companions. Even if one member of a binary cannot be seen, its mass can be estimated from the orbit of the visible star, as deduced from

● Above left: *A black hole, the final stage in the life of a massive star. At the hole's centre is a singularity, the remains of the former star crushed to a point of infinite density. Around the singularity is the 'surface' of the hole, an invisible sphere within which gravity is so great that not even light can escape. The surface of the black hole, known as the event horizon, lies at the so-called Schwarzschild radius. Light entering this region is heavily reddened by the gravitational red shift before it disappears. Outside the Schwarzschild radius, the black hole's gravity bends passing light beams, as though by a lens, but does not capture them.*

● Above: *One of the most likely candidates for a black hole is an X-ray source called Cygnus X-1, here seen from the surface of a hypothetical orbiting asteroid. What is believed to be the black hole is accompanied by a blue supergiant from which gas is being drawn off. Spiralling around the hole, the gas heats up and emits X-rays before finally falling into the hole.*

spectroscopic observations. If the companion is a black hole, gas from the visible star will be drawn towards it like water towards a plughole. The gas spirals around the hole before finally vanishing into it, heating up to millions of degrees and emitting X-rays that can be detected by satellites above the Earth's atmosphere.

Astronomers have detected a number of high-mass X-ray sources in binary systems which could be black holes. The most celebrated is Cygnus X-1, which consists of an X-ray source with an estimated mass of around eight Suns orbiting a visible blue supergiant in the constellation Cygnus. Another likely black hole binary is the X-ray source prosaically labelled A0621−0 in the constellation Monoceros. But the most massive black holes, containing the equivalent of millions of Suns, are believed to lie at the centres of galaxies and quasars, and may account for the violent activity taking place far off in the Universe, as we shall see in the final chapter.

THE INNER PLANETS

3

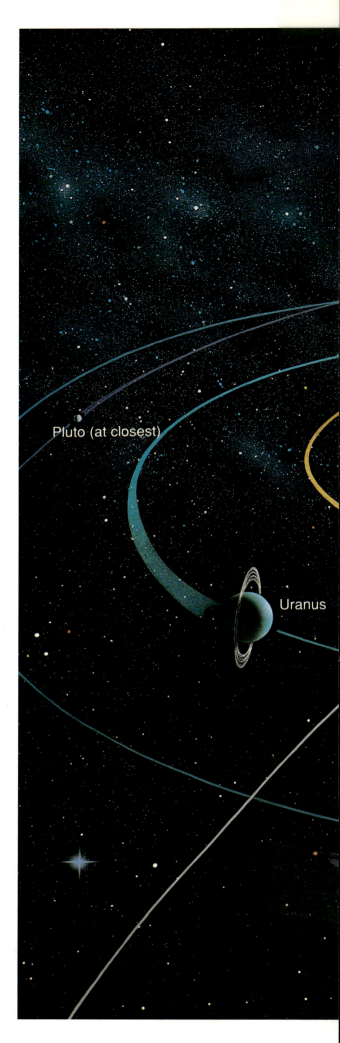

The Greeks called them *asteres planetai*, 'wandering stars', but they are not stars at all. Telescopes, and more recently space probes, have shown us what the early skywatchers could never have known – planets are worlds, each with its own character, from small rocky bodies to gaseous giants. Their 'wandering' occurs because they are in orbit around the Sun, and so is really not a wandering but a precisely predictable motion governed by the laws of gravity. Planets emit no light of their own; they shine by reflecting sunlight.

The roll-call of the planets in order of distance from the Sun is: Mercury, Venus, Earth, Mars, Jupiter, Saturn, Uranus, Neptune, and Pluto. The last on the list, Pluto, is an oddity, for its orbit partially overlaps Neptune's. These nine planets, together with their moons and assorted debris such as asteroids and comets, make up the Solar System. The only planets without moons are the two innermost ones, Mercury and Venus. If we could observe the Solar System from afar and get the kind of view shown here, our overriding impression would be one of flatness – the orbits of the planets lie in almost the same plane, with only comets to disturb the overall picture.

The origin of the Solar System, including the creation of planet Earth, is understood in general terms, and is believed to be intimately linked with the formation of the system's central pivot, the Sun. The dating of rocks from the Earth, the Moon, and meteorites, which is done by measuring the relative amounts of particular radioactive atoms they contain, shows them to be around 4600 million years old. That was when the Sun and planets took shape.

Within the cosmic fog of the nebula that spawned it, the embryonic Sun was surrounded by a spinning disk of surplus material, both gas and dust. Gradually, through random collisions, the dust specks began to stick together, building up progressively larger bodies ranging in size from small stones to blocks as big as mountains. This process is called *accretion*. Once mini-planets a few kilometres across had formed, their gravity was sufficient to attract other bodies, accelerating accretion and building up a family of planets, many with moons in orbit around them.

Near the proto-Sun, where temperatures were highest, only small, rocky bodies could form – the inner planets, from Mercury to Mars. But in the cooler outer reaches of the Solar System the planetary cores could gather around them a thick cloak of gas, building up the giant planets

● *Overview of the Solar System (not to scale), showing the nine planets from Mercury to Pluto. Pluto is shown at its closest to the Sun (a position it reached in 1989), when it comes within the orbit of Neptune. A comet with a long flowing tail is depicted on its elongated orbit. Jupiter, Saturn, Uranus, and Neptune all have rings, but only Saturn's are bright enough to be seen from Earth.*

Pluto (at closest)

Uranus

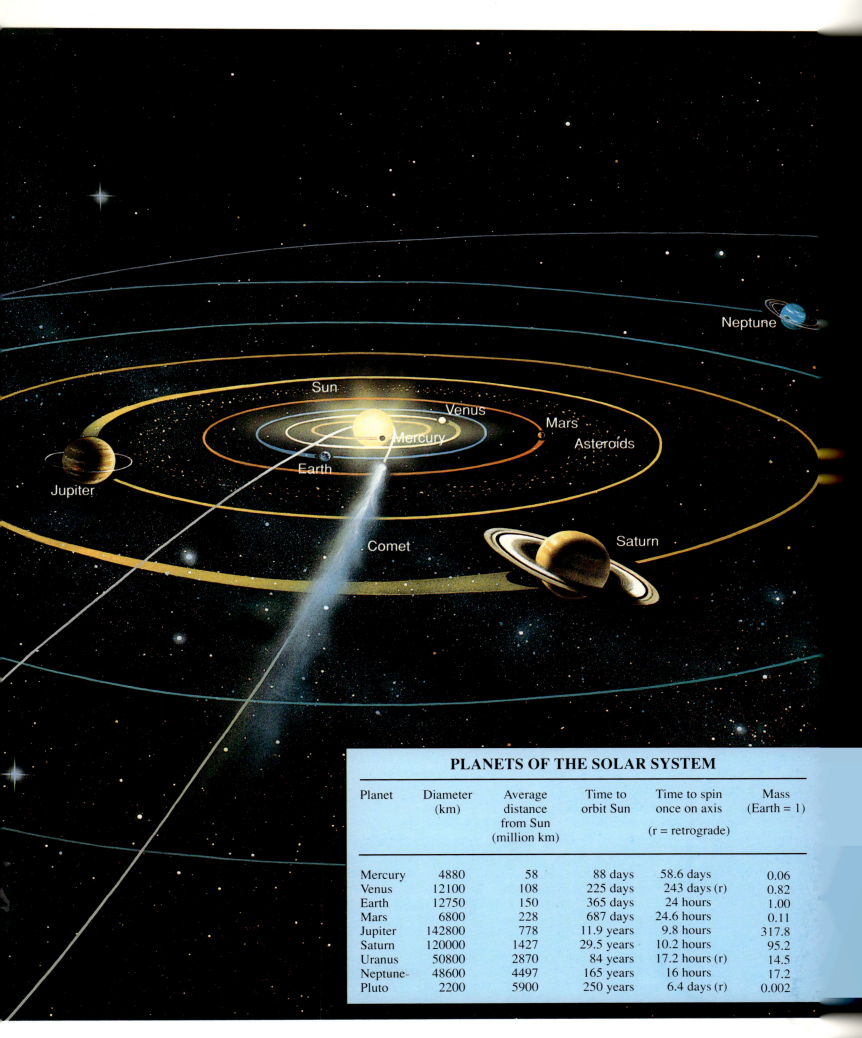

PLANETS OF THE SOLAR SYSTEM

Planet	Diameter (km)	Average distance from Sun (million km)	Time to orbit Sun	Time to spin once on axis (r = retrograde)	Mass (Earth = 1)
Mercury	4880	58	88 days	58.6 days	0.06
Venus	12100	108	225 days	243 days (r)	0.82
Earth	12750	150	365 days	24 hours	1.00
Mars	6800	228	687 days	24.6 hours	0.11
Jupiter	142800	778	11.9 years	9.8 hours	317.8
Saturn	120000	1427	29.5 years	10.2 hours	95.2
Uranus	50800	2870	84 years	17.2 hours (r)	14.5
Neptune	48600	4497	165 years	16 hours	17.2
Pluto	2200	5900	250 years	6.4 days (r)	0.002

from Jupiter to Neptune. The outermost planet, Pluto, is little more than a punctuation mark separating the planets from the realm of the comets.

While all this planetary construction was going on, the Sun was still glowing gently from heat released by its contraction. The onset of nuclear reactions within the infant Sun gave rise to an energetic solar wind that drove away the remaining gas and dust, cleaning up the Solar System and leaving it much as we see it today. Remaining chunks of debris collided with the planets and their moons during an era of intense bombardment that lasted for several hundred million years and left enduring scars.

● *How the Solar System is thought to have been born. From a disk of matter surrounding the young Sun, progressively larger bodies built up (top left). When the Sun's nuclear reactions switched on, the resulting solar wind of atomic particles blew surplus gas and dust out of the Solar System along with the primitive atmospheres of the inner planets (centre). At bottom right is the inner Solar System as we see it today, with Earth and Moon in the foreground.*

What evidence is there to support this theoretical picture, apart from the existence of the Solar System itself? Astronomers know of a class of objects called T Tauri stars which appear to be young stars swaddled in a cloud of gas and dust that may form the raw material for a future planetary system. Our Sun is believed to have passed through such a stage shortly after its nuclear reactions switched on.

More significantly, in 1983 a satellite called IRAS (the Infrared Astronomy Satellite) discovered clouds of gas and dust around forty of the nearest stars including the bright star Vega, which is considerably younger than the Sun. According to the IRAS measurements the material around Vega consists of solid chunks ranging from small pebbles to objects many kilometres across, exactly what is expected for a planetary system in the early stages of formation.

Another of the IRAS sources is the star Beta Pictoris, an otherwise unremarkable star in the southern hemisphere. In 1984, American astronomers photographed the material around Beta Pictoris and found it to be spread into a disk two or three times the diameter of our Solar System, similar to the disk from which our own planetary system is believed to have formed. Presumably planet formation is still going on around Beta Pictoris and Vega, as well as the other stars on the IRAS list.

These findings suggest that not only do other planetary systems exist, but they are relatively common throughout the Galaxy. Fully formed planets around other stars would be too faint to be seen through Earth-based telescopes, although there is hope that telescopes in space, like the Hubble Space Telescope, may detect them.

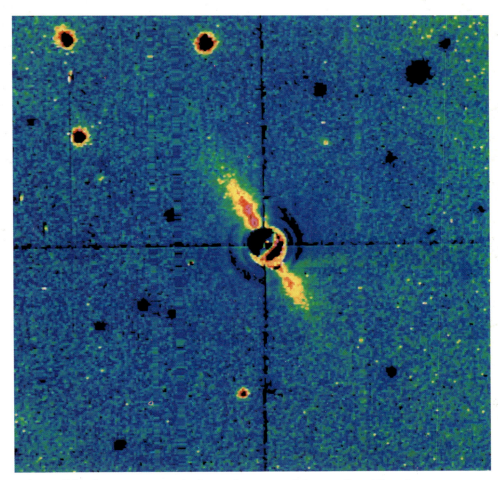

● *A possible planetary system in formation around the star Beta Pictoris, seen on an artificially coloured-coded picture. The material forming the planetary system is a disk presented edge-on and appears as an elongated red and yellow streak. To get this picture, light from the star itself was blotted out by a circular mask supported on cross-wires.*

● *According to the ancient Greek view of the Universe (top), the Earth was at the centre and around it revolved the Sun and planets, here shown figuratively by Andreas Cellarius in his atlas of 1660 (see also p.14). In the view of Copernicus (above), first published in 1543, the Sun was at the centre, and the Earth merely a planet orbiting it.*

N. COPERNICUS.

● Right: *Nicolaus Copernicus, the Polish astronomer who in 1543 dethroned the Earth from its position at the centre of the Universe.*

Closest of the planets to the Sun is tiny Mercury, orbiting the Sun in only 88 days and named after the fleet-footed messenger of the gods. It never strays far from the Sun in the sky and, although at its best it can appear as bright as Sirius, it is difficult to spot because it is usually lost in twilight, low in the sky, particularly as seen from higher northern latitudes. From the northern hemisphere, Mercury is best seen in the evening sky in the spring, and in the morning sky in the autumn. Binoculars reveal it as a bright star, somewhat orange in colour, and once found in binoculars it may be possible to locate it with the naked eye. Mercury is easiest to see when it is at its widest angular separation from the Sun (known as *greatest elongation*).

If we could travel to the surface of Mercury, the Sun would appear much larger than it does from Earth, and its size would change noticeably because Mercury's orbit is distinctly elliptical, varying between 46 million and 70 million km (29 million and 43 million miles) from the Sun. When Mercury is at its closest (*perihelion*) the Sun would seem three times as large as seen from Earth, but when Mercury is at its most distant (*aphelion*) the Sun would appear only twice the size as seen from Earth.

Mercury is a midget – only 4880 km (3030 miles) in diameter, two-fifths of the Earth's size and smaller than any other planet except Pluto; both Jupiter and Saturn have moons that are larger than Mercury. Because of its small size and its proximity to the Sun, Mercury cannot hold an atmosphere. It is a totally arid and airless ball of rock, blasted by solar radiation. Its daytime side is roasted to over 400°C at perihelion, hot enough to melt lead, while at night the temperature drops to an unimaginably frigid −170°C. Certainly, this is not a world that astronauts would want to visit.

Seen through a telescope, Mercury goes through phases as it orbits the Sun, like the phases of the Moon. Beyond that, it is difficult to make out much detail on its surface, apart from the occasional dusky patch. For many years astronomers believed that the planet's rotation period was the same as its orbital period, which would mean that Mercury kept one face permanently turned towards the Sun, in the same way that the Moon keeps one face turned towards the Earth. On this assumption, several astronomers drew maps of Mercury that showed vague patterns of light and dark markings. However, the maps are worthless because the assumption was wrong.

● *Mercury and Venus are easiest to see around the time of greatest elongation from the Sun, when they are at their maximum angular separation either east of the Sun (in the evening sky) or west of it (in the morning sky). At superior and inferior conjunctions Mercury and Venus are in line with the Sun and hence lost from view.*

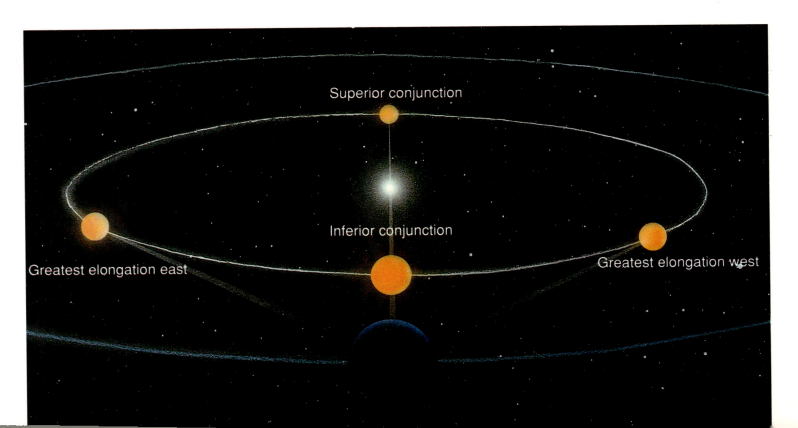

Superior conjunction

Inferior conjunction

Greatest elongation east

Greatest elongation west

The embarrassing truth emerged in 1965, when some American astronomers bounced radio waves off the planet's surface, a technique known as radar astronomy. From the change in frequency of the returned signal they could tell how quickly the planet rotated. It turned out to be not 88 days after all, but exactly two-thirds of that, 58.6 days. So for every two orbits of the Sun, Mercury spins three times on its axis, a cycle lasting 176 days. This is also the time taken for the Sun to return to the same place in Mercury's sky, for instance from one noon to the next.

It seemed likely that Mercury, not much bigger than our Moon, would look like the Moon. This guess was confirmed in 1974 when the US space probe Mariner 10 flew past the planet, photographing almost half its surface from close range. At first glance it is difficult to tell a photograph of Mercury from one of the Moon. Craters of all sizes abound on Mercury, most of them formed by the impact of debris left over from the birth of the Solar System, the newest being surrounded by bright rays of material splashed out by the impact that formed them.

● Below left: *Mercury, photographed by the Mariner 10 space probe, is covered with impact craters including some with bright rays that give it a resemblance to our own Moon.*

● Below: *Close-up of Mercury showing some of the long, low scarps that wrinkle the surface. The large crater with a central peak at bottom right, named Gluck, is 85 km (53 miles) across.*

● Below right: *Mercury occasionally passes across the face of the Sun, an event known as a transit. Here it is seen as a tiny black dot near the Sun's limb at the transit of 1986 November 13.*

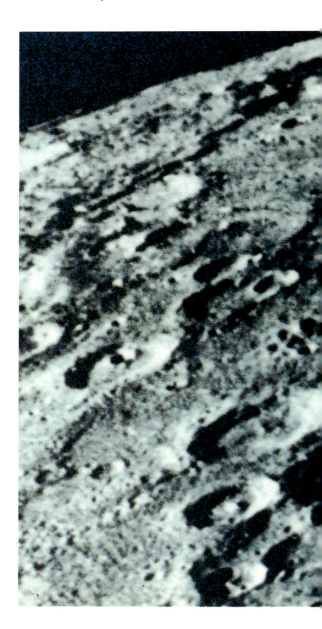

The largest feature on the known portion of Mercury is a huge circular plain called the Caloris Basin, 1300 km (800 miles) across and ringed by mountains 2 km (1.2 miles) high. Similar in size to the Mare Imbrium on the Moon, the Caloris Basin was evidently blasted out by an asteroid of considerable size about 4000 million years ago. Completion of the mapping of Mercury's surface is a job for future spacecraft.

Although Mercury has a rocky surface, the planet's high average density means that it cannot be composed of rock all the way through. Much of its interior must consist of iron – indeed, it is believed to have an iron core that extends for fully three-quarters of its diameter, proportionally far larger than the iron core of the Earth or any other planet. How did Mercury come to have such an unusually large core? One possibility is that it once had a thicker rocky crust that was knocked off in a gigantic collision with a smaller body early in the history of the Solar System, at a time when such destructive impacts are thought to have been common.

When Mercury was young this iron core was probably liquid. As it cooled and solidified it contracted, so that the whole planet shrank slightly. This shrinking caused the crust to wrinkle, and those wrinkles are visible today as long scarps, no more than a few kilometres high but extending for hundreds of kilometres across the surface.

From time to time Mercury crosses the face of the Sun, an event known as a *transit*. Transits can happen only when the planet lies between the Earth and the Sun (*inferior conjunction*), in May and November, for those are the months when Mercury crosses the plane of the Earth's orbit. Since Mercury's orbit is tilted at 7° to the Earth's orbit, the inner planet will pass either above or below the disk of the Sun at other times.

During a transit Mercury appears as a tiny black dot that takes several hours to crawl across the Sun's face. The event can be followed by projecting the Sun's image as when observing sunspots (described in Chapter Two). At first the disk of Mercury might be mistaken for a particularly circular sunspot, but it gives itself away by its steady movement, and in comparison with Mercury's jet-black silhouette sunspots appear distinctly brownish. Mercury will next transit the Sun in 1993 on November 6, in 1999 on November 15, in 2003 on May 7, and in 2006 on November 8.

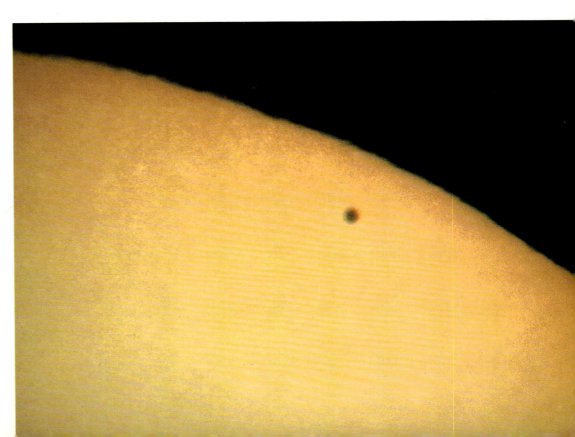

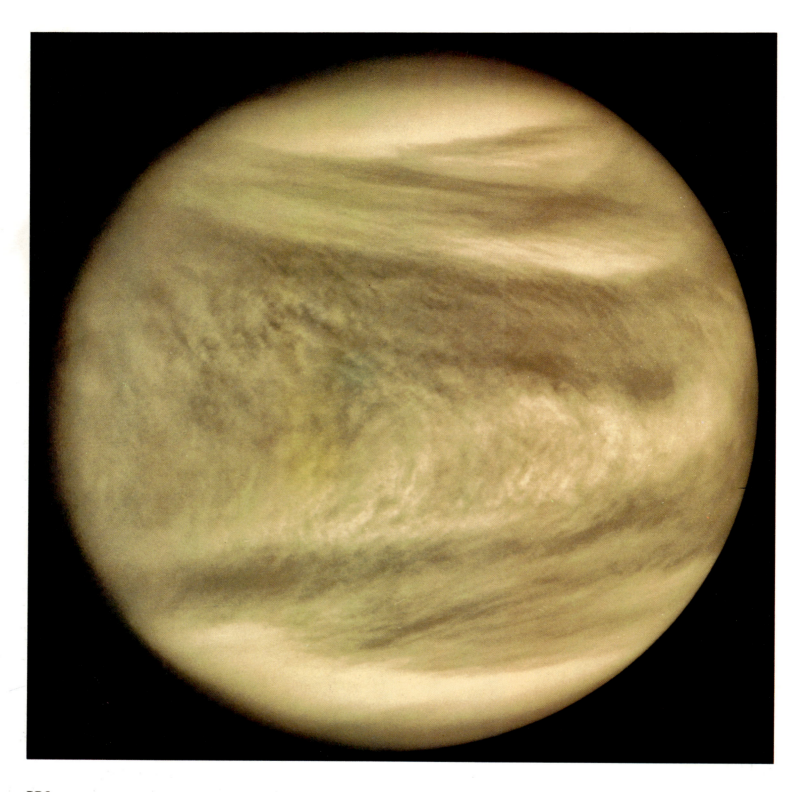

Whereas Mercury is difficult to see, Venus is impossible to miss. It appears brighter than any other planet, and at its best shines 20 times more brightly than the star Sirius. Everyone has seen Venus at one time or another, even if they have not recognized it, for it is the brilliant evening or morning 'star', first to appear in the west after sunset or last to vanish in the brightening east before sunrise.

The main reason for its brilliance is that Venus is shrouded in unbroken clouds which reflect most of the sunlight hitting them. For all its brilliance, there is little to see on Venus through a telescope because the clouds permanently hide its surface from view. A telescope will show the planet's phases as it orbits the Sun every seven-and-a-half months. When Venus is a narrow crescent it is relatively close to the Earth and appears so large that the phase can be seen in binoculars. Some people have even seen Venus as a crescent with the naked eye, although the feat requires exceptional eyesight. At its closest, when passing between the Earth and the Sun, Venus can come to within 40 million km (25 million miles) of us, closer than any other planet.

Venus is almost the same size as the Earth, a mere 650 km (400 miles) smaller in diameter, and science fiction writers once speculated that under its clouds Venus would turn out to be a twin of the Earth in other ways too. The truth is less romantic, but hardly less astounding. Venus, in fact, is an incarnation of Hell.

Starting with the American Mariner 2 that flew past it in 1962, Venus has been thoroughly investigated by space probes from the USA and the USSR. Under its clouds Venus is intensely hot, with an average temperature of around 470°C both day and night, similar to the hottest parts of the planet Mercury, and with little difference between equator and poles. Venus has suffered a runaway version of the greenhouse effect that is currently thought to be warming our own planet.

The main reason for the high temperature is that the atmosphere consists almost entirely of carbon dioxide, the gas released by burning fossil fuels on Earth (and exhaled by humans). The carbon dioxide traps heat like a highly efficient blanket, equalizing the temperatures around the planet, and bearing down at the surface with a pressure of 90 Earth atmospheres – enough to flatten all but the most heavily reinforced spacecraft.

Why is there so much carbon dioxide in the atmosphere of Venus? To answer this question, we must look at what has happened to the Earth's atmosphere. Surprisingly, there was once as much carbon dioxide in the atmosphere of Earth as there now is on Venus. The difference is that on Earth the gas was absorbed in the oceans and eventually became laid down in rocks such as limestone. Carbon dioxide is broken down by plants, which retain the carbon for their bodies and release oxygen for us to breathe. Venus has no oceans (indeed, it has scarcely any water at all) and no life, so the carbon dioxide remains in its atmosphere to drive up the temperature.

However, carbon dioxide on its own is not sufficient to explain the extreme temperature of Venus. The remaining cracks in the greenhouse are sealed by the small amount of water that exists in the upper atmosphere, and the planet's clouds. These clouds are unlike anything on Earth. Whereas terrestrial clouds are composed of water vapour, the uppermost clouds of Venus consist of droplets of concentrated sulphuric acid, stronger than in a car battery. From the clouds falls a corrosive acid rain.

What's more, they are found at a much higher altitude than terrestrial clouds. On Venus the main cloud deck is about 50 km (30 miles) up, with two thinner layers of haze above it at about 58 and 65 km (36 and 40 miles). By comparison, the highest Earth clouds, cirrus, are at heights of about 10 km (6 miles). Winds of up to 350 kph (220 mph) whip the uppermost layer of clouds around Venus in four days, far more quickly than the planet itself spins.

It was not until the first radar observations of the planet's surface in the early 1960s that we had any idea of what the rotation period of Venus actually was. There was a double surprise – not only does Venus spin far more slowly than any other planet, but it does so from east to west, the opposite way from the other planets. (This is known as *retrograde* rotation.) The rotation period of Venus is 243 Earth days, longer than the 225 days the planet takes to orbit the Sun. Even now there is no generally accepted explanation of this slow, reverse rotation. The Sun's gravity may well have had an effect, but Venus may also have suffered a huge impact during or shortly after its formation that knocked its rotation backwards.

It is sometimes said that the 'day' on Venus is 243 Earth-days long, but this is misleading. If a day is defined as the time from one fixed point

● *The clouds of Venus, photographed in ultraviolet light by the Pioneer-Venus Orbiter probe. The clouds spiral away from the mottled equatorial region towards the brighter poles, forming V-shaped markings. The upper clouds, composed mostly of strong sulphuric acid, circulate around the planet every four days.*

such as sunrise or noon to the next, then the day on Venus lasts 117 Earth days, which is the planet's rotation period relative to the Sun (although of course the Sun is never visible from the permanently cloud-bound surface). The clouds, incidentally, rotate in the same direction as the planet spins, from east to west. They spiral around Venus from pole to equator, creating V-shaped markings that show up clearly on space probe photographs taken through ultraviolet filters.

Although we cannot see the surface, radio waves can penetrate the clouds and so it has been possible to map Venus by radar, both from spacecraft and from large radio telescopes on Earth. The resulting maps show two main continents: the larger, called Aphrodite Terra, lies just south of the equator and is half the size of Africa; the other, Ishtar Terra, in the northern hemisphere, is the size of Australia. Ishtar is distinguished by the highest mountains on Venus, called Maxwell Montes, which soar to almost 11 km (7 miles) – higher than Mount Everest – above the planet's average surface level (the equivalent of 'sea level'). Another upland area, called Beta Regio, appears to be a volcanic ridge 2500 km (1500 miles) long and 5 km (3 miles) high.

Most of the surface is covered with rolling plains, apparently similar to the lowland areas on the Moon known as maria. Lava-flows hundreds of kilometres long extend across the plains, some issuing from craters and others from fissures. Some of those volcanoes are still active today, releasing the sulphur that goes to form the uppermost clouds of Venus. One eruption may have occurred in the late 1970s, when astronomers recorded a temporary surge in the sulphur content of the planet's upper atmosphere. Not all the craters are volcanic, though. Numerous impact craters, some over 100 km (60 miles) in diameter, have been detected – even the dense atmosphere of Venus does not block meteorites 1 km (0.6 miles) or more across.

Soviet spacecraft have actually landed on the surface and sent back photographs showing a rock-strewn landscape. Chemical analyses carried out by the spacecraft showed that the rocks of Venus are similar to terrestrial volcanic basalt. About as much sunlight penetrates to the surface as on a cloudy day on Earth, although the light is a sulphurous orange. Unlike the high-speed winds that prevail in the upper atmosphere, at the surface the wind speed is a sluggish 3 or 4 kph (2 mph). The lander craft had to be highly reinforced and refrigerated to survive the hostile conditions at the surface of Venus. Even so, they lasted less than an hour before succumbing. The exploration of Venus will be best left to robots, not humans.

● Right: *Venus beneath its clouds, as seen by radar aboard the Pioneer-Venus Orbiter. Heights are colour coded and range from 2.9 km (1.8 miles) below the average surface level (equivalent to sea level) at the lowest point to 10.8 km (6.7 miles) above average at the highest, Maxwell Montes in the northern hemisphere. Most of Venus is covered by rolling plains but there are two main continents, Ishtar Terra in the northern hemisphere, shown by yellow and brown contours, and Aphrodite Terra near the equator. There are large volcanoes and some impact craters on Venus. This Mercator projection distorts the size of features near the poles, making them seem larger than they really are.*

● Below: *The rocky surface of Venus, bathed in an eerie orange glow under the dense clouds, photographed by the Soviet Venera 14 lander. At the bottom can be seen part of the lander itself, with a serrated edge. The curved object on the surface is a discarded lens cover, and at right is a colour chart.*

Transits of Venus across the Sun are much rarer than for Mercury, occurring in pairs more than a century apart. They were once an important way of finding the scale of the Solar System, for the difference in the positions of Venus against the Sun as measured from two widely separated locations on Earth would allow the planet's distance to be calculated by simple trigonometry. Captain Cook's explorations of New Zealand and Australia were undertaken during an expedition to observe the transit of 1769 from Tahiti. Nowadays more accurate measurements are used, such as radar ranging. The last transits of Venus were in 1874 and 1882, and the next will be in 2004, on June 7, and in 2012, on June 5.

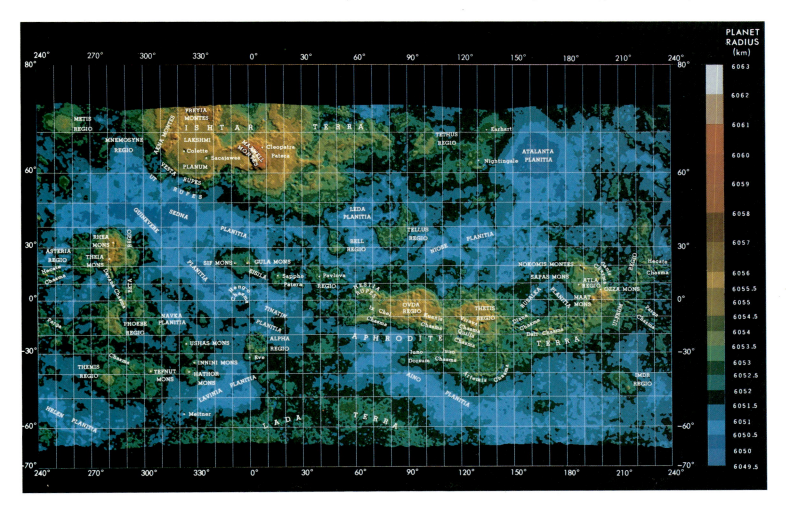

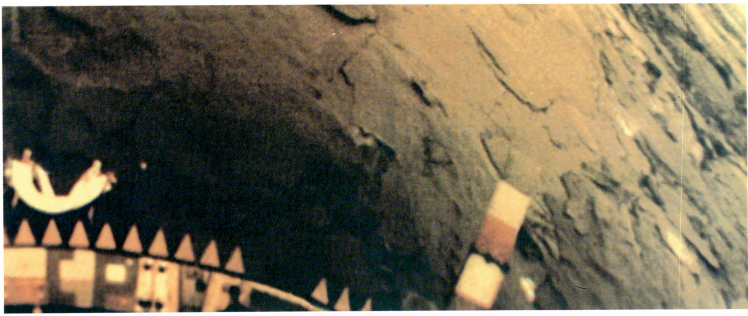

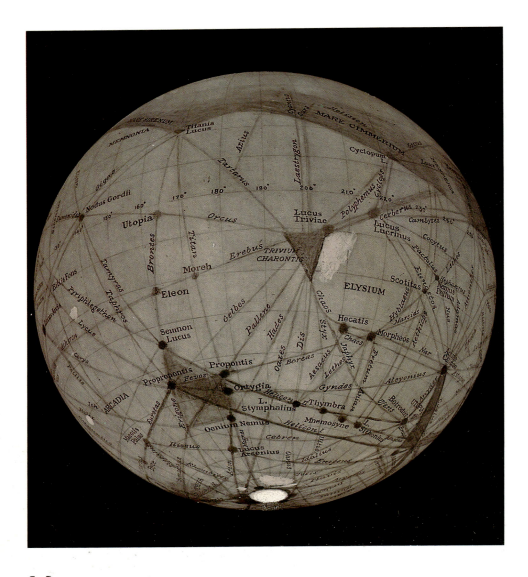

● Left: *A globe of Mars showing the network of canals which American astronomer Percival Lowell believed had been dug by Martian beings to irrigate their crops. Note the 'oases' where canals cross. The canals are now known to have been illusory.*

● Above: *Percival Lowell, relaxing at the Lowell Observatory.*

● Above right: *Orson Welles, whose radio broadcast of* War of the Worlds *created panic among listeners who believed that the Earth really was being invaded by Martians.*

Mars, named after the god of war, has an angry red glint that becomes particularly noticeable when it is at its nearest and brightest. More than any other planet, Mars has inspired dreams of alien life and now inspires dreams of exploration and colonization.

The planet bears a superficial resemblance to the Earth. Its day is only 40 minutes longer and its axial tilt only slightly more (25.2° against 23.4°) so that it has seasons like ours. Visible through a telescope are clouds that come and go, and icy-white polar caps that shrink and advance with the changing seasons.

These apparent similarities with Earth seriously misled some astronomers, notably a rich American amateur named Percival Lowell who established his own observatory at Flagstaff, Arizona, specifically to study the red planet. Lowell was inspired by the work of Giovanni Schiaparelli, an Italian, who at the close approach of Mars in 1877 had detected linear markings that he termed *canali*, meaning channels. Schiaparelli thought they might possibly be rivers carrying water from the melting polar caps, but Lowell interpreted them literally as 'canals', which he believed to be the work of intelligent Martians.

At the subsequent close approach of Mars in 1894, Lowell began a series of observations lasting many years during which he mapped a planetwide network of canals. He expounded his ideas of a Martian civilization in books such as *Mars as the Abode of Life*, vividly describing the canal system he believed the Martians had dug from pole to equator to irrigate their drought-ridden planet. Each canal, he suggested, was flanked by a fertile strip – like that along the River Nile – which made the canals more noticeable from Earth, even broadening into rounded oases where two canals crossed.

However, other astronomers failed to see the canals. Under the best observing conditions the canals broke up into irregular streaks and spots of such complexity that it was impossible to draw them all accurately. It was easy to understand how the eye could join these disconnected details into imaginary straight lines where none actually existed. The respected planetary observer Eugene Antoniadi spoke for the sceptics in 1930 when he dismissed Lowell's canals as "completely illusory," attributable to tricks of the eye. Opposition to Lowell came not just from astronomers but also from the naturalist Alfred Russel Wallace, a colleague of Charles Darwin, who in 1907 pointed out that Mars was too cold and dry for any kind of advanced life.

Part of the problem is that only occasionally does Mars come close enough to Earth to be well seen. As Mars orbits the Sun it passes the Earth every 26 months, an event known as *opposition* (see illustration), but because its orbit is markedly elliptical, its distance from Earth at opposition varies widely. At the most distant oppositions, as in 1995, it is 100 million km (60 million miles) away, nearly twice as far as at its closest, as it was in 1988. The closest approaches occur only every 15 or 17 years. However, this ellipticity of the orbit of Mars played an important part in astronomical history, for it allowed Johann Kepler to deduce the first of his laws of planetary motion (see the box on p. 72).

● *The distance of Mars at different oppositions varies noticeably because of the planet's elliptical orbit.*

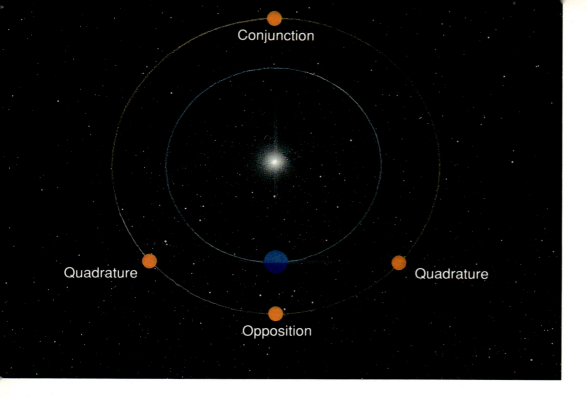

Conjunction

Quadrature

Quadrature

Opposition

● Left: *The outer planets come closest to the Earth at opposition, when they are opposite in the sky from the Sun and so appear due south at midnight. At quadrature they are 90° from the Sun (when Mars shows a noticeable phase); at conjunction they are lost from view on the far side of the Sun.*

● Below: *Tycho Brahe, at right, prepares to make an observation with an enormous sighting instrument mounted on a wall, while an assistant below holds a candle and another prepares to record the observation. The wall is decorated with a picture of Tycho and shows other instruments in his observatory on the island of Hveen.*

Mars is just over half the diameter of the Earth, but the land area of the two planets is virtually identical because there are no seas on Mars, whereas two-thirds of the Earth's surface is covered with water. The red deserts of Mars are marked by darker patches that change in shape from season to season, and from year to year, so it is understandable that astronomers believed they might be areas of vegetation, even after the more extreme Lowellian theories had been disposed of. Moss or lichen were the usual candidates, because only they seemed hardy enough to survive the harsh conditions on the planet.

Those who still expected to see signs of civilization on Mars, or indeed of life of any kind, were sorely disappointed by the photographs taken by the Mariner 4 space probe in 1965, which revealed large impact craters like smoother versions of those on the Moon. This impression of a lunar-like Mars was reinforced four years later by Mariners 6 and 7, which photographed more craters and found that the atmosphere is composed of carbon dioxide with a surface pressure less than a hundredth of the Earth's. Of canals there was no sign, and although the pictures were not good enough to show up anything living it was clear that the conditions were even more hostile than had been feared.

THE ORBIT OF MARS AND KEPLER'S LAWS

The true shape of the planets' orbits was discovered early in the seventeenth century by a German mathematician, Johann Kepler, an assistant to the Danish astronomer Tycho Brahe. Tycho, the greatest observer of the pre-telescopic era, had painstakingly measured the positions of the planets over a period of twenty years from his observatory on the island of Hveen, between Denmark and Sweden. Kepler found that Tycho's data for Mars could not be explained by a circular orbit. In fact, of the outer naked-eye planets, Mars is the one whose orbit departs most noticeably from a circle, but it took Tycho's skilled observations to reveal this fact.

After six years of patient calculation Kepler established that the true shape of the orbit of Mars, and the orbits of all the other planets, are ellipses, a finding enshrined as his first law of planetary motion. Kepler's first law was published in 1609 along with the second law, which states that the speed of a planet varies with its distance from the Sun, being fastest when the planet is closest (at perihelion). A third law, relating the period of a planet's orbit to its distance from the Sun, was published ten years later. Kepler's laws established the correctness of the Sun-centred theory of Copernicus, and paved the way for Newton's theory of gravitation which explained why planets move the way they do.

● *Johann Kepler, the German mathematician who discovered that the orbits of the planets are elliptical.*

● *Mariner Valley, a huge rift valley 4500 km (2800 miles) long, runs horizontally across this mosaic of Mars made from images taken by the Viking 1 orbiter. At the left is a chain of three huge volcanoes, Ascraeus Mons (top), Pavonis Mons, and Arsia Mons. The largest volcano on Mars, Olympus Mons, lies off the picture to the left. Dried-up river valleys run northwards from Mariner Valley towards the dark Mare Acidalium at top right. Between Mariner Valley and Mare Acidalium lies Chryse, the lowland area where the first Viking lander touched down. The white patches are thin cloud.*

So far, the Mariner probes had only snatched quick looks at parts of the planet as they flew past. What was needed was a complete reconnaissance of Mars from orbit, a task that was accomplished with spectacular success by Mariner 9 during a year-long mission starting in 1971. There, missed by previous probes, were volcanoes to dwarf anything on Earth. The largest of them, which corresponds to a circular feature visible through telescopes, is named Olympus Mons (Mount Olympus), after the home of the mythical Greek gods. It is the largest volcano in the Solar System, 600 km (400 miles) broad and 25 km (15 miles) high with a complex summit crater where clouds often form on spring and summer afternoons. Wrinkled lava-flows cover the surrounding plain, suggesting that Olympus Mons has been active within the past few hundred million years – relatively recent by geological (and astronomical) standards.

● Below: *Olympus Mons, the largest volcano on Mars, rises above a foaming sea of white cloud in this Viking orbiter view.*

Another circular feature seen from Earth, Hellas, proved to be an immense lava-filled impact basin 1500 km (950 miles) wide. Its floor is the lowest point on Mars, about 5 km (3 miles) deep. Morning mists form here as they do in an immense rift valley, Valles Marineris, that stretches for 4500 km (2800 miles). Valles Marineris, initially formed by faulting of the crust and subsequently eroded by winds, corresponds to a broad, dark 'canal' visible from Earth called Coprates.

Perhaps most exciting of all was the discovery of what appear to be dried-up river beds on the surface of Mars – evidence of climatic change. Several explanations have been advanced, but it seems most likely that early in its history Mars had a denser atmosphere than it does today, consisting of carbon dioxide and water vapour released by erupting volcanoes. Temperatures and pressures were then high enough for liquid water to exist on the surface. Had all the water been released at once it could have formed a planet-wide ocean 50 metres (150 feet) deep. In practice, though, the water was probably released over a billion years or more, and – as in any desert – it did not collect into pools but quickly evaporated or soaked into the ground. Over time the carbon dioxide was incorporated into the surface rocks, the global temperature fell, and Mars became locked in the grip of an ice age. Much of the water became frozen into a subsurface permafrost layer. Additional water is stored in the polar caps, mixed with carbon dioxide ('dry ice') that freezes out of the atmosphere in winter.

Mariner 9 finally solved the mystery of why the darker features on Mars appear to change in intensity and shape from year to year, changes that were once attributed to the growth and dying-back of vegetation. The answer is dust, either covering or uncovering areas of darker rock as it is blown by seasonal winds. There is no shortage of dust on Mars: sometimes huge dust storms envelop the whole planet, as they did in 1971 when Mariner 9 arrived. Most dark markings are due to slight differences in surface composition rather than actual physical features. The most prominent of all the dark areas on Mars, the wedge-shaped Syrtis Major, is a slope that descends gently to the west of a cratered highland region.

Paradoxically, although Mariner 9 showed previous speculations about life on Mars to be unfounded, the knowledge that liquid water existed there in the past raised new hopes that some form of life may once have arisen – not little green men, certainly, but perhaps little green bugs. Two space probes called Viking landed on the red planet in 1976 specifically to look for life.

● Below right: *This Viking mosaic shows Mangala Valley, looking like a dried-up river bed, presumably produced by running water when the climate of Mars was milder than today.*

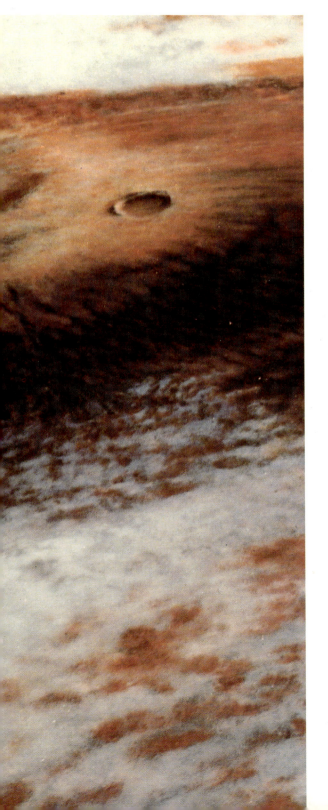

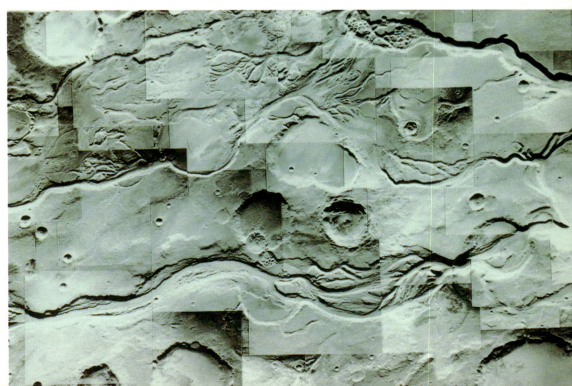

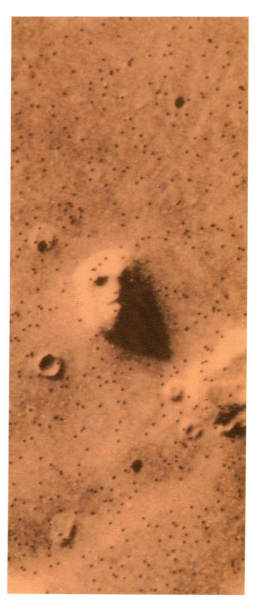

Through the eyes of Viking, we know what it is like to stand on Mars. A rock-strewn desert stretches to the horizon, which is broken by the outlines of craters. The sky above is pink, because of fine dust particles suspended in the atmosphere. Winds are light for much of the time, and the daytime temperature remains permanently below freezing, even on a summer's afternoon, dropping to −100°C or lower before dawn. The rocks get their redness from a coating of iron oxide, better known on Earth as rust, produced when oxygen from the atmosphere combined with iron in the soil. Mars is a planet that has literally rusted away.

Each Viking carried a miniature biology laboratory in which samples of Martian soil, scooped up by a mechanical hand, were incubated in three different ways in the hope of making any microorganisms grow. The results told us a fair amount about the physical and chemical nature of the Martian soil, but as for finding life they were negative. Another experiment, which analysed the soil in search of molecules from the bodies of any organisms, alive or dead, also drew a blank.

● Far left: *The end of another day – sunset over the red sands of Mars viewed from the Viking 1 lander in the planet's northern hemisphere. Part of the lander is just visible at the bottom right.*

● Left: *The 'face' on Mars. Shadows on an eroded mountain 1.5 km (1 mile) across give the impression of a human head in this highly enlarged Viking orbiter picture.*

In retrospect, the lack of any life on Mars is not too surprising for the atmosphere is so thin that lethal amounts of ultraviolet radiation from the Sun penetrate to the surface. When humans land on Mars, they will find a planet as inimical to life as the Moon. Mars bases would have to be buried under the surface to protect the occupants from radiation, meteorites, and the scouring effect of dust storms driven by winds of up to 300 kph (200 mph).

Mars has two satellites, Phobos ('fear') and Deimos ('terror'), named after the mythical sons of the god of war who accompanied him on the battlefield. Small and faint, they are difficult to see and were not discovered until the close approach of 1877, the same year in which Schiaparelli sighted the *canali*. Their discoverer was Asaph Hall of the US Naval Observatory in Washington, DC, who had the advantage of using the Observatory's new 66 cm (26 inch) refractor. He also had the advantage of a wife who urged him to continue searching when he was ready to give up after several fruitless nights. Thus encouraged, he returned to the task and found both moons in the course of one week.

Phobos, the closer and larger of the two, orbits a mere 6000 km (3700 miles) above the surface of the planet. Its orbital period is 7 hours 40 minutes, less than a third of the planet's axial rotation period, so to an observer on Mars, Phobos would appear to go across the sky the opposite way to all other celestial bodies, rising in the west and setting in the east five-and-a-half hours later. Deimos orbits 20,000 km (12,500 miles) above the planet's surface every one-and-a-quarter days. Tidal forces are causing Phobos to spiral in towards Mars. At the current rate of progress it would crash to the surface in about 40 million years, but before then it will break up to form a ring of debris around the planet.

Space probes have given us close-up views of these irregular-shaped moons, which look rather like lumpy potatoes. Phobos has an average diameter of only 22 km (14 miles), while Deimos is a mere 12 km (7.5 miles) across. Both are probably former asteroids that strayed too close to Mars and were captured by its gravity, and both are dented by impact craters. The second-largest crater on Phobos is named Hall, after the moon's discoverer; the largest is named Stickney, the maiden name of his wife.

● Left: *The rocky red desert of Mars, as seen from the Viking 1 lander. Even the sky is reddish, as a result of fine dust in the atmosphere. On the right are trenches dug by Viking's soil sampler. The white arm carried instruments to measure wind-speeds, and air temperature and pressure. Left of centre is a large rock 2 metres (6 feet) across, nicknamed Big Joe.*

● Right: *Phobos, the larger of the two moons of Mars, is an irregular object with a large crater, Stickney, 10 km (6 miles) across, from which radiate grooves. The impact that formed Stickney nearly shattered the tiny moon. Phobos is probably a captured asteroid.*

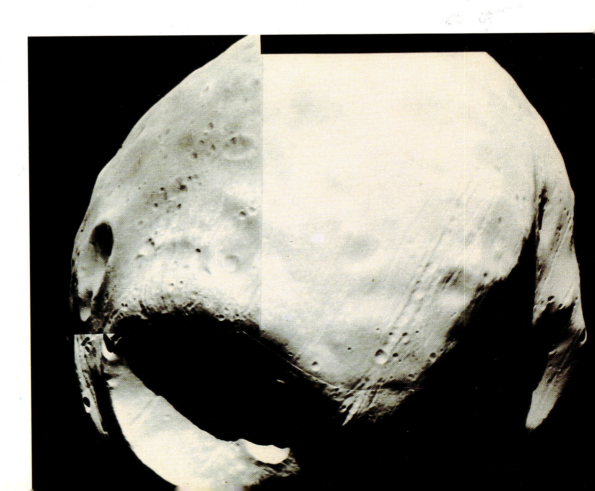

Between the orbits of Mars and Jupiter lies a gap, a suspiciously large gap as Johann Bode pointed out in 1772. Towards the end of the eighteenth century astronomers began to wonder whether one or more small planets might exist unseen in this region of the Solar System. In 1800 a Hungarian astronomer, Baron Franz Xaver von Zach, decided to organize a network he called the Celestial Police to track down such bodies. He issued invitations to 24 selected astronomers to join the task force, each of them to be assigned a section of sky.

One of those von Zach decided to approach was Giuseppe Piazzi of Palermo Observatory, Sicily. But before Piazzi had even received his invitation he spotted a moving 'star' in the constellation Taurus in the early hours of January 1, 1801. Further observation showed that it was orbiting the Sun between Mars and Jupiter, and hence appeared to be the kind of object the Celestial Police were looking for. It was named Ceres, after the patron goddess of Sicily.

In 1802 the German astronomer Heinrich Olbers discovered a second such body, Pallas. A third, Juno, was found by Karl Harding in 1804, and Olbers added a fourth, Vesta, in 1807. They are known as minor planets or, more popularly, *asteroids*. Since the advent of photography thousands of asteroids have been discovered, and more are being added all the time. Baron von Zach himself never did discover an asteroid, although asteroid no. 999 is named after him. Asteroid no. 1000 is named in honour of Piazzi.

Ceres is by far the largest asteroid, with a diameter of about 1000 km (600 miles) but, surprisingly perhaps, it is not the brightest. That honour goes to Vesta, which is just over half the diameter of Ceres but has a much lighter-coloured surface. At its best Vesta is just detectable with the naked eye, but all the first four asteroids to be discovered are bright enough to be seen easily in binoculars.

● *Giuseppe Piazzi, the Sicilian astronomer who discovered the first and largest asteroid, Ceres, in 1801.*

● *Most asteroids orbit in a belt between the planets Mars and Jupiter, but some stray farther afield. Here an asteroid passes beneath the southern polar cap of Mars. The two moons of Mars are thought to be captured asteroids.*

Asteroids are thought to be the leftovers from the formation of the Solar System, and there are estimated to be 100,000 of them larger than 1 km (0.6 miles) orbiting in a belt between Mars and Jupiter. Even if they were all put together they would make a body far smaller than our Moon.

Certain asteroids lie outside the main belt, including a family that moves along the same orbit as Jupiter. These lie in two groups, 60° ahead and behind the planet, at gravitationally stable locations termed *Lagrangian points*. They are named after heroes of the Trojan wars, and so are known as the Trojan asteroids. The outermost of Jupiter's 16 moons are believed to be captured asteroids.

Perhaps the oddest asteroid of all is Chiron, discovered in 1977, which lies even further from the asteroid belt than the Trojans. Its eccentric 50 year orbit carries it from almost as far as the orbit of Uranus to within the orbit of Saturn. It now seems that Chiron is actually a large cometary nucleus, 180 km (110 miles) in diameter, and is similar in many ways to Saturn's outermost moon, Phoebe.

Of particular concern to us are the asteroids that cross the Earth's orbit, since they may collide with us – indeed, one such collision has been blamed for the death of the dinosaurs (see p. 139). The first asteroid of this kind was discovered in 1932 and named Apollo; consequently, any member of this family is termed an Apollo asteroid. Some of them may be the remains of 'dead' comets.

One Apollo asteroid, known simply as 1989 FC, passed 690,000 km (430,000 miles) from the Earth in March 1989, less than twice the distance of the Moon and closer than any other known body; it was not seen until a week before its closest approach. The previous near-miss record was held by asteroid Hermes, which came to within 800,000 km (500,000 miles) of the Earth in 1937.

At first, asteroids were named after characters from mythology, but as the number of discoveries piled up the available names rapidly ran out. Nowadays whoever finds an asteroid is allowed to name it, which accounts for some curious appellations such as Swissair (apparently the discoverer's favourite airline), Mr Spock (the character in *Star Trek*), The NORC (a computer), and various relatives and girlfriends. Four asteroids have been named in honour of the Beatles.

4 THE OUTER PLANETS

On the night of January 7, 1610, Galileo Galilei, professor of mathematics in the northern Italian town of Padua, was scanning the heavens through his newest and most powerful telescope, made from spectacle lenses. Jupiter blazed high and bright among the stars of Taurus. Galileo had seen the planet before through less powerful telescopes, and knew that it showed a disk – unlike the stars, which remained points of light. But on this night he noticed something new: "Three little stars, small but very bright, were near the planet," he wrote afterwards. The following night he saw them again, but this time arranged differently.

As he watched the planet over the next few nights his suspicions about these "little stars" were confirmed. They were not stars at all, but moons, moving around Jupiter in their own orbits like planets around the Sun. On January 13 he discovered a fourth. They are now collectively known as the Galilean satellites.

Jupiter is the largest planet in the Solar System, eleven times the diameter of the Earth, and the first of four giants far from the Sun, totally different in nature from the Earth and its fellow rocky inner worlds. Jupiter is a ball of liquid wrapped in a mantle of gas. Yet despite the fact that it is composed mostly of the lightest elements in the Universe, hydrogen and helium – similar to the composition of the Sun, in fact – it is so large that all the other planets rolled together would not weigh half as much as Jupiter.

An ordinary pair of binoculars will show Jupiter as clearly as did Galileo's primitive telescope, and only a modest telescope is needed to see detail on the planet's face. The first thing that becomes apparent is that Jupiter is crossed by alternating bands of light and dark cloud like veins in marble, the most prominent of these being two dark belts north and south of the equator. More careful inspection reveals that the disk of the planet is not perfectly round, but is distinctly flattened at the poles. The cause in both cases is the swift rotation of Jupiter, which spins on its axis in less than 10 hours, drawing the clouds out into belts parallel to the equator and distorting the planet's shape. The equatorial regions rotate 5 minutes faster than the rest of the planet.

Jupiter's stormy clouds contain a variety of chemicals, giving them a range of colours – mostly shades of yellow, red, and brown, with tinges of blue. If we were to parachute into the clouds, as a space probe called

● *A landscape of fire and brimstone – volcanoes spray out sulphur on Jupiter's hot moon Io, turning its surface orange. Io is the most volcanically active body in the Solar System.*

(appropriately) Galileo is due to do in December 1995, we would first pass through the upper cloud layer, experiencing a temperature of around −150°C and a pressure about half that at the Earth's surface (i.e. half an atmosphere). At a depth of about 75 km (45 miles) the temperature reaches 0°C and the pressure has risen to several atmospheres. Rain and sleet may exist here, along with lightning and hurricane-force winds that shake our descending capsule. At a depth of 1000 km the atmospheric pressure is so great that hydrogen is squeezed into a hot liquid. Waves may roll languidly over this liquid hydrogen sea, but the nearest thing that the planet has to surface features are in its cloud layers.

The space probes Pioneer 10 and 11 passed the planet in 1973 and 1974, and Voyagers 1 and 2 followed in 1979. The images they transmitted to Earth revealed convection patterns in the planet's clouds: the white zones are areas of ascending gas, topped by high cirrus clouds of ammonia, while the dark belts are areas where the gas descends again. As the descending air moves either polewards or equatorwards, Coriolis forces generated by the planet's swift rotation – 45,000 kph (30,000 mph) at the equator – produce contraflowing winds in neighbouring belts and zones that blow east and west at hundreds of kilometres per hour.

The darker, lower-altitude belts probably get their colour from sulphur and tarry molecules containing carbon. On the night side of Jupiter the Voyagers photographed immense splashes of lightning whose energy may help create the complex chemicals that tone the planet's clouds.

Sitting like a Cyclopean eye in the planet's southern hemisphere is the highest cloud feature of all, the Great Red Spot. First seen in 1664, it is the nearest thing to a permanent feature on Jupiter, remaining constant in latitude but drifting back and forth in longitude. It is certainly great, being long enough to swallow three Earths, but it is not always red – often it appears pink or grey, and at times it disappears entirely, leaving a colourless hollow. Even now the nature of the Great Red Spot is not fully understood, but it appears to be a spinning storm cloud. Its colour is believed to come from red phosphorus welling up from below, the amount of phosphorus governing the spot's varying intensity.

● Above: *Galileo Galilei, the Italian scientist who discovered the four main moons of Jupiter in January 1610, and a page from his book* Sidereus Nuncius, *in which he reported his discovery of their nightly movements.*

● Left: *Jupiter and its volcanic moon Io photographed by Voyager 1. Io can be seen at centre right of Jupiter's disk, against the planet's swirling clouds. The Great Red Spot is visible towards the bottom left.*

Other spots and storms appear from time to time in the planet's clouds. Some white ovals just south of the Great Red Spot have lasted over 50 years, while other features have had much shorter lifetimes. The key to the storminess of Jupiter's clouds is that the planet is still giving off heat left over from its formation, which stirs up the overlying layers. Without this internal heat source the weather on Jupiter would be very dull.

As we have seen, Jupiter's atmosphere is not much deeper than the Earth's, but beneath it lies a sea of liquid hydrogen about 24,000 km (15,000 miles) deep. Deeper still, the pressures become so great – three million atmospheres and more – that the hydrogen molecules are broken into free protons and electrons. Hydrogen in such a state is electrically conductive, like a metal – but it is not solid for it is too hot, with a temperature of 10,000°C or more. This so-called metallic hydrogen region is 40,000 km (25,000 miles) deep.

● Right: *Jupiter's Great Red Spot in close-up. The spot is a rotating storm feature in Jupiter's clouds. Several times larger than the Earth, it has persisted for several centuries, fed by heat from Jupiter's interior. Festoons of cloud swirl around the red spot. Just south of it is an oval white cloud.*

● Below: *Colourful cloud bands on Jupiter. The colours, produced by various chemicals in the planet's atmosphere, have been enhanced by image processing in order to bring out detail.*

At the very centre of the planet is thought to lie a rocky core only 50 per cent larger than the Earth, but with at least ten times the Earth's mass. Here the temperature is 30,000°C, several times hotter than the surface of the Sun. In fact, had Jupiter been born with 50 times as much mass it would have been hot enough at its centre for nuclear reactions to switch on, turning it into a small star.

Convection currents in Jupiter's hot metallic hydrogen interior generate a magnetic field, probably in the same way that the Earth's magnetic field is produced in its iron core. The difference is that in the case of Jupiter the generator is vastly bigger and so, consequently, is the magnetic field. Were it visible to the naked eye, the magnetic field around Jupiter would appear larger than the full Moon. Within this magnetic wrapping orbit the four Galilean satellites.

Atomic particles are trapped in the magnetic field to produce a more intense version of the Earth's Van Allen radiation belts. Future space explorers would do well to steer clear for fear of being subjected to lethal doses of radiation. The magnetic axis of Jupiter is tilted at about 11° to the planet's rotation axis, similar to the Earth's magnetic tilt, but the north and south magnetic poles are the opposite way round to the Earth's.

Around Jupiter orbits a family of 16 moons like a mini Solar System, some of which are as interesting as the planets themselves. Indeed Jupiter's largest moon, Ganymede, is larger than Mercury, and Callisto is only slightly smaller. Through the television eyes of the Voyager probes, Galileo's "little stars" became worlds in their own right.

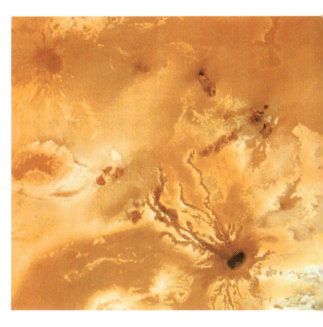

Ganymede turns out to be the largest moon in the Solar System, 5260 km (3270 miles) across, just pipping Saturn's main attendant, Titan. Ganymede has an icy crust, the oldest parts of which have been coated by dark dust; the largest dark area is named Galileo Regio. The most unusual features are strange grooves presumably caused by crustal movements. Bright impact craters star its surface, but not as abundantly as on Callisto, which is the most heavily cratered body in the Solar System.

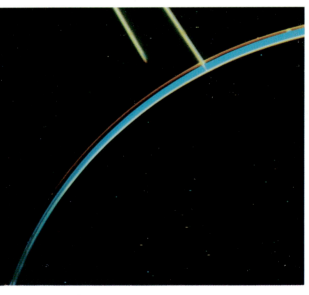

● Top left: *Cracks in the icy crust of Jupiter's moon Europa.*
● Left, above left: *Sulphur volcanoes on Io.*
● Top: *The heavily cratered surface of Callisto.*
● Right: *Faults on Ganymede.*
● Above: *Jupiter's faint, dusty ring (top) illuminated by sunlight from behind the crescent planet.*

Callisto's surface could not be more cratered, since every available part of it bears the scars of impacts, many of them obliterating older features below. Its largest feature is a basin called Valhalla, 3000 km (2000 miles) across, reminiscent of the Caloris Basin on Mercury or Hellas on Mars. On Callisto, crustal movements have not obliterated craters or disturbed the dark dust that gives it the darkest surface of the Galilean satellites.

The brightest surface belongs to Europa, encased in an egg-like shell of ice that is as reflective as the clouds of Venus. Its surface is veined with cracks through which water is believed to have flowed from below, spreading out over the surface and filling in any craters before freezing to form an almost perfectly smooth, uniform surface.

Undoubtedly the most spectacular of all Jupiter's moons – indeed, of all moons in the Solar System – is the innermost of the Galilean quartet, Io. Although scarcely larger than our own Moon, Io is more volcanically active than the Earth. When the Voyager probes flew past in 1979 they spotted eight volcanoes erupting, spraying sulphur and sulphur dioxide to heights of up to 300 km (200 miles). The ejected sulphur paints Io's landscape various shades of orange, while the sulphur dioxide falls to form bright white patches.

In all, hundreds of volcanic vents puncture the moon's surface, from some of which oozes liquid sulphur. They are named after mythical gods of fire such as Prometheus, Loki, and Pele. Unlike the volcanoes of Earth, these vents do not sit atop mountains but open directly onto the surface; Io is in fact remarkably flat. This fire-and-brimstone body is endlessly recycling its outer few kilometres, resurfacing itself to a depth of about 1 metre (3 feet) every 10,000 years. At this rate of eruption any impact craters are soon filled in.

Why is Io's interior so hot? There are two likely causes. Firstly it is being continually kneaded by the gravitational pulls of Jupiter and the moons Europa and Ganymede. (Europa's watery interior has been partially melted by the same effect.) A second source of energy is electrical currents induced in the moon as it moves rapidly through Jupiter's magnetic field.

Another discovery by the Voyagers was a gossamer-thin ring of fine dust particles girdling Jupiter's equator, too insubstantial to see from Earth. The ring, about 6000 km (4000 miles) broad, lies 50,000 km (30,000 miles) above the planet's cloud tops and is made up of dust from two tiny moonlets that orbit at its outer edge. Apart from the Galilean satellites, all Jupiter's moons are small, and the outer ones are believed to be captured asteroids.

● Right: *In a remarkable odyssey 12 years and seven billion kilometres long, Voyager 2 passed Jupiter, Saturn, and Uranus before finally reaching Neptune, at the edge of the Solar System. From here the probe's radio messages, travelling at the speed of light, took over four hours to reach Earth. Its sister ship, Voyager 1, encountered Jupiter and Saturn, but in passing Saturn's moon Titan its path was bent steeply upwards so that it could not visit Uranus and Neptune.*

● Below: *The movement of Jupiter's four main moons over a period of four hours. Ganymede moves in front of the planet from right to left (its shadow is seen on the planet in the top picture). Below it, Io moves away, passing under Callisto, which is moving towards Jupiter, while Europa remains almost stationary at the far right. Such movements can easily be followed with small telescopes or even binoculars.*

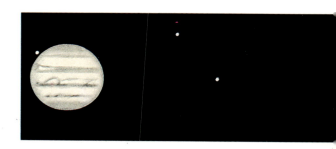

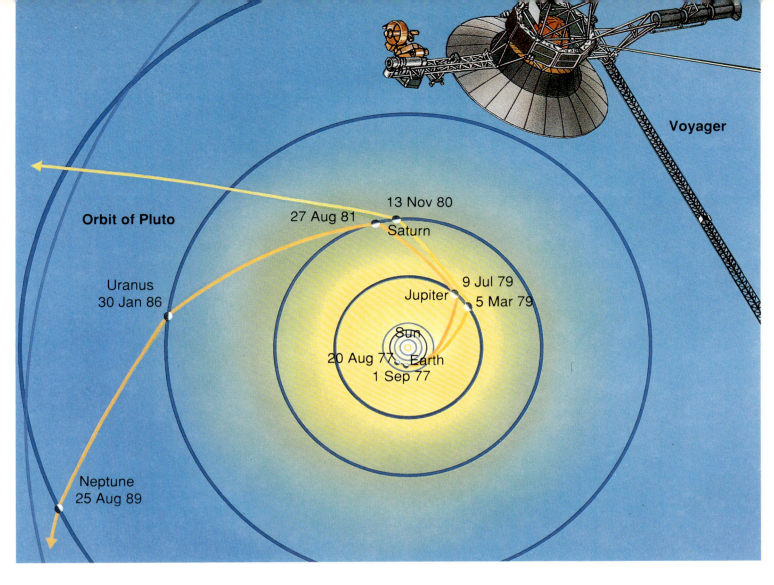

Voyager

Orbit of Pluto

13 Nov 80
27 Aug 81
Saturn

Uranus
30 Jan 86

9 Jul 79
Jupiter
5 Mar 79

Sun
20 Aug 77   Earth
1 Sep 77

Neptune
25 Aug 89

VOYAGER'S INTERSTELLAR MESSAGE

Neptune was Voyager 2's last port of call before it headed off into the Galaxy, in a different direction from Voyager 1. Just in case any aliens should one day chance upon these robot emissaries and wonder about their origins, both probes carry a plaque containing engravings of humans and information about the location of the Sun. There is also a long-playing record that contains spoken greetings in various languages, a selection of Earth sounds, and 90 minutes of music ranging from Beethoven to Chuck Berry. Also coded into the grooves are more than a hundred pictures showing the Earth and life on it. Ground controllers expect to maintain contact with the Voyagers until about the year 2020. However, it will be millions of years before the Voyagers come close to another star.

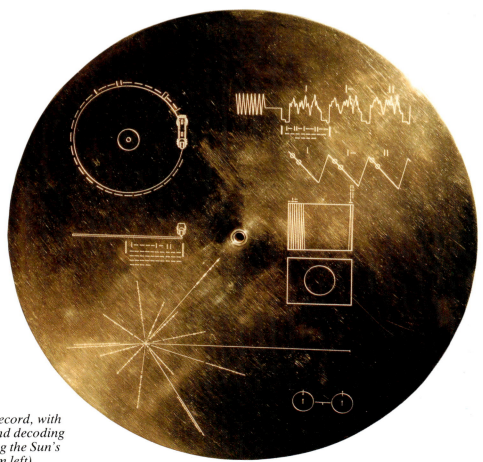

● Right: *Cover of the long-playing Voyager record, with instructions for playing the record (top left) and decoding pictures from it (top right), and a map showing the Sun's location with respect to several pulsars (bottom left).*

In Greek mythology, Cronus overthrew his father Uranus and reigned over Heaven. But in an attempt to forestall a prophecy that he too would be overthrown in like fashion, he devoured his children as they were born. Cronus is better known by his Roman name, Saturn, after whom the outermost planet visible to the naked eye is named. When Galileo observed Saturn in the summer of 1610 he thought it had two large moons. But his primitive telescope was not good enough for him to be certain of what he was seeing, so he cautiously published his impressions as an anagram in Latin:

SMAISMRMILMEPOETALEUMIBUNENVGTTAVRIAS

Decoded and translated, this announcement reads: "I have observed the highest [i.e. most distant] planet to be triple." Not long after, though, the two companions vanished. "Does Saturn devour his children?" queried a puzzled Galileo, mindful of the mythological tale.

Not until over 40 years later did telescopes improve sufficiently for astronomers to see the reason for these baffling changes in the appearance of Saturn. A Dutch astronomer, Christiaan Huygens, realized that the planet is in fact surrounded by a bright ring. In the course of Saturn's stately progress around the Sun in its 29½ year orbit, the rings are presented to us at different orientations: sometimes they are tilted towards us at a steep angle, when they are most easily seen, while at other times they are edge-on and temporarily vanish from view. It was these changing aspects of the rings that baffled Galileo and subsequent observers. Forthcoming passages through the plane of Saturn's rings will occur in 1995 and 1996, and 2009.

Huygens reasoned that the rings could not be solid, but must be made up of a swarm of moonlets in orbit around the planet. This interpretation was confirmed 20 years later at Paris Observatory by Giovanni Cassini, who discovered a gap in the rings, now known as the Cassini Division, the width of the Atlantic Ocean.

● *Medal showing Christiaan Huygens, the Dutch astronomer who realized that Saturn was encircled by a ring.*

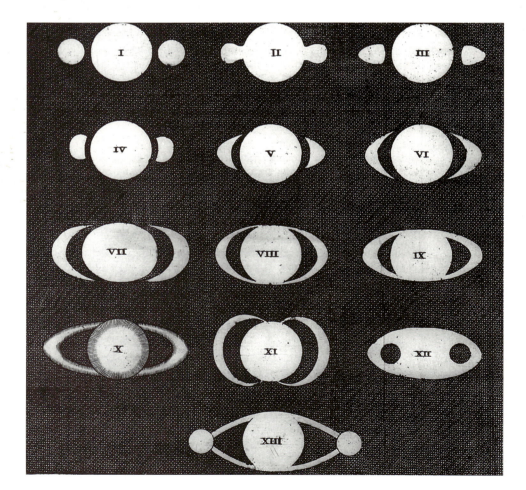

● *Interpretations of the appearance of Saturn by various early observers, published in Christiaan Huygens' book* Systema Saturnium.

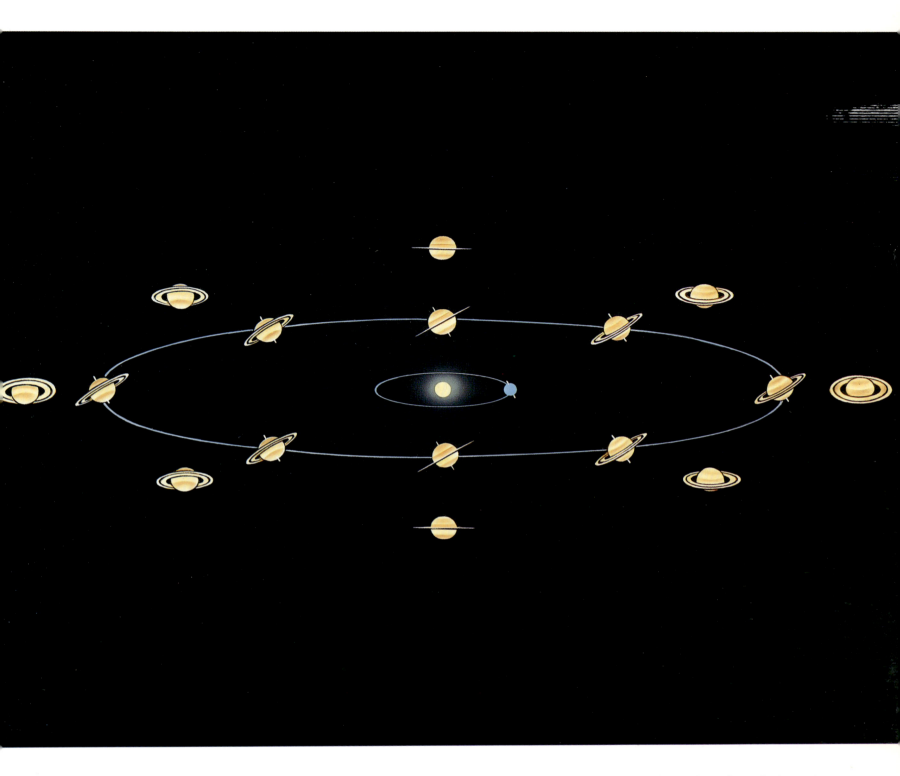

● *As Saturn orbits the Sun its rings are presented to us on the Earth at various angles, causing changes in the planet's appearance that fooled early observers.*

In many ways Saturn is a scaled-down version of Jupiter, being of similar composition (i.e. hydrogen and helium) but nine times the diameter of the Earth as against Jupiter's eleven times. However, Saturn has by far the lowest density of all the planets – lower even than the density of water. Outwardly, this low density makes itself apparent in the planet's outline, which is far more squashed than that of any other planet, its polar diameter being fully 10 per cent less than its equatorial diameter. Of course, it is only the *average* density that is this low, for Saturn is much denser than water at its centre. Like Jupiter, the planet probably has a rocky core surrounded by a layer of liquid metallic hydrogen, but not as deep as Jupiter's because Saturn has a lower mass and hence lower internal pressures.

Saturn is another fast rotator, spinning once every 10¼ hours at the equator but a little more slowly towards the poles. A telescope shows an ochre disk with subdued cloud bands. There is no equivalent of the

Great Red Spot, and only occasionally do markings of any kind become prominent to observers on Earth. Voyagers 1 and 2 photographed some swirling storms and spots when they flew past the planet in 1980 and 1981, but overall the clouds are far less turbulent and colourful than those of Jupiter. This is surprising because Saturn gives out at least as much internal heat as Jupiter. One suggested explanation is that high-altitude haze masks atmospheric activity from view, but it may be that Saturn simply behaves differently, for reasons not understood. For example, there are strong easterly winds that blow around the equator at speeds of up to 1800 kph (1100 mph), four times faster than on Jupiter. A joint US–European space probe called Cassini will find out more when it reaches the planet early next century.

The distinguishing feature of this most elegant of planets is its system of bright, equator-circling rings. They stretch for 275,000 km (170,000 miles) from rim to rim, yet are no more than a few hundred metres thick. In relation to their width, the rings of Saturn are thinner than a sheet of tissue paper, which is why they vanish when we see them edge-on.

● Right: *Saturn's placid ochre disk and encircling rings are a calm contrast to Jupiter's febrile, garish cloudscape. The moons Tethys, Dione, and Rhea are visible in this Voyager picture; the dark dot on the planet is Tethys's shadow. The shadow of the planet falls on the rings.*

● Below: *Details in the clouds of Saturn can be brought out by image processing, as in this Voyager view of the planet's northern hemisphere. In general, though, the clouds of Saturn are much less active and colourful than Jupiter's.*

There are three main components to the rings. Ring A is the outermost, separated by the dark Cassini Division from Ring B, which is the densest and hence the brightest of the three. The innermost and faintest part is called Ring C, also known as the Crepe Ring because it is transparent and the body of the planet can be seen through it. The rings are composed of snowballs, ranging in size from specks to large boulders, and mutual collisions are continually grinding them down. Orbital periods of the bodies in the rings range from about 6 hours at the inner edge of Ring C to 14 hours in Ring A.

When photographed from close range by the Voyager probes the rings broke up into innumerable threadlike ringlets, giving the appearance of the ridges and grooves on an enormous gramophone record. Narrow ringlets were found even in the Cassini Division. Overlying the bright Ring B are dark blotches, known as *spokes*, which come and go over periods of a few hours. The spokes are believed to consist of fine dust suspended above the rings by static electricity, generated by interactions between the rings and Saturn's magnetic field.

If the rings could be swept together they would make a modest-sized moon a few hundred kilometres across. Most likely they consist of material that was too close to Saturn to form into a satellite;

● Right: *The strangely kinked F ring of Saturn, seemingly composed of two intertwined strands. The paths of the F ring particles are controlled by two 'shepherd' moons orbiting either side of the ring.*

● Below right: *Saturn and its moons Tethys and Dione. The shadow of the rings falls on Saturn's cloud tops, and the body of the planet can be seen through the broad Cassini Division in the rings.*

● Below: *shadowy radial features called spokes, overlying the threadlike rings of Saturn, photographed by Voyager 2.*

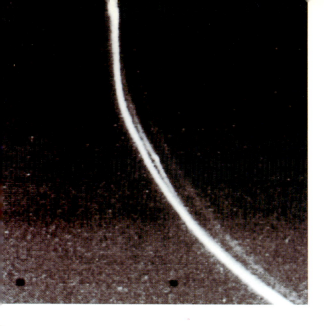

alternatively, they may be the debris from a former satellite that strayed too close to the planet and broke up under its tidal force.

Some 4000 km (2500 miles) beyond the rim of Ring A lies the slender F ring. This is the most baffling ring of all, for in places it seems to split like a frayed strand of cotton. Either side of it orbit two small moons, discovered by Voyager 1 and named Pandora and Prometheus. These flanking moons, each about 100 km (60 miles) across, are termed *shepherds* because, like sheepdogs, they prevent the particles of the F Ring from straying. However, their gravitational pull also produces a kinked appearance in the ring. An even smaller moon, Atlas, orbits 800 km (500 miles) outside Ring A and keeps its outer edge trim. These new moons around Saturn, discovered by the Voyagers, bring the known total to at least eighteen, more than around any other planet, with several more 'probables' and 'possibles' not yet confirmed. All Saturn's moons consist of a mixture of ice and rock.

By far the largest of the moons is Titan, discovered in 1655 by Huygens. It is the only moon in the Solar System to have a substantial atmosphere – in fact, the atmospheric pressure at the surface is over half as much again as on Earth. Perhaps the greatest surprise of all is that Titan is the only body other than the Earth with an atmosphere predominantly of nitrogen. Methane is also present in small quantities, and it may condense into lakes. Unfortunately we cannot see the surface because it is veiled by a permanent smog of orange cloud. Titan itself has a diameter of 5150 km (3200 miles), making it slightly smaller than Jupiter's moon Ganymede, but if the thick cloud layer is taken into account the diameter is about 5550 km (3450 miles), making it bigger than Ganymede.

Although Titan is far too cold for life, scientists think that some of the first chemical steps to life may be taking place in its dense atmosphere. We could look upon this moon as a deep-freeze version of the Earth, nearly ten times as far from the Sun. In 2002, when the Cassini space probe is due to go into orbit around Saturn, a smaller sub-probe called Huygens will parachute down to the surface of Titan to investigate its icy crust.

Among Saturn's other satellites, perhaps the most unusual is Iapetus, black on one side and icy white on the other. Iapetus is the outermost but one of Saturn's moons, and its harlequin nature might be due to an accumulation of dust knocked off a neighbouring moon, or to its having swept up material left over after its formation.

Mimas, a moon just under 400 km (250 miles) in diameter, is scarred by a crater fully one-third its own diameter; the crater is named Herschel, after the moon's discoverer, and the impact that formed it must almost have broken Mimas apart. Such a fate apparently befell the predecessor of two small, irregular bodies called Epimetheus and Janus that now move in near-identical orbits just beyond the rings of Saturn.

In two other cases, the same orbit is used by more than one of Saturn's satellites. Dione, 1100 km (700 miles) across, has a small body preceding it in the same orbit, while Tethys, virtually the same size as Dione, shares its orbit with two smaller fragments, one orbiting ahead of it and one behind.

Until 1781, the Solar System was thought to end at Saturn. In that year a musician and amateur astronomer, William Herschel, caused one of the greatest sensations in the history of science when he discovered a new planet whose existence had been totally unsuspected. The planet was twice as far from the Sun as Saturn, so his discovery literally doubled the size of the known Solar System overnight.

It happened while the 42-year-old Herschel was systematically surveying the sky from the back garden of his house in the English town of Bath. Herschel, German-born, had come to England to play and teach music, but found himself increasingly consumed by astronomy. He built his own telescopes, and it was with one of these instruments with a mirror 15 cm (6 inches) wide that he was observing the stars on this fateful night.

As he described it later: "On Tuesday the 13th of March, between ten and eleven in the evening, while I was examining the small stars in the neighbourhood of H Geminorum, I perceived one that appeared visibly larger than the rest: Being struck with its uncommon magnitude, I compared it to H Geminorum and the small star in the quartile between Auriga and Gemini, and finding it so much larger than either of them, suspected it to be a comet."

He referred to it as a comet, since neither he nor anyone else was thinking in terms of a new planet, but soon Herschel and others who observed the object realized that it did not look like a comet at all. Its outline was round and sharp, unlike the fuzzy, elongated shapes of comets, and it did not move in the protracted orbit typical of a comet. After it had been observed for some months, long enough for its orbit to be computed, astronomers finally concluded that they were looking at a new planet.

Herschel was now a celebrity. King George III appointed him court astronomer, and in return Herschel tried to have the planet named after the king. Other astronomers at first called it Herschel, but eventually tradition prevailed and it was named Uranus, after the mythological father of Saturn.

Uranus turned out to be another gaseous planet, not as big as Jupiter or Saturn but still four times the diameter of the Earth. Although it is so remote, nearly 3000 million km (1800 million miles) from the Sun, it is easily seen in binoculars and at best is just visible to the naked eye to someone who knows exactly where to look. However, even through a powerful telescope Uranus displays little more than a featureless greenish disk, with none of the cloud banding of Jupiter and Saturn.

● *William Herschel, the amateur astronomer who discovered the planet Uranus in 1781 and was subsequently sponsored by the king of England, George III.*

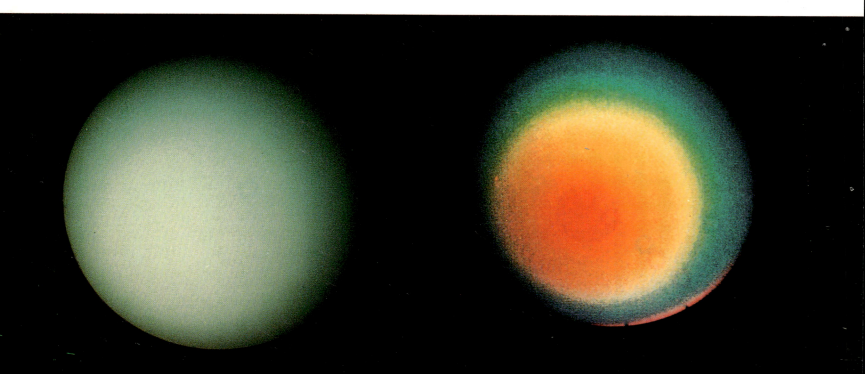

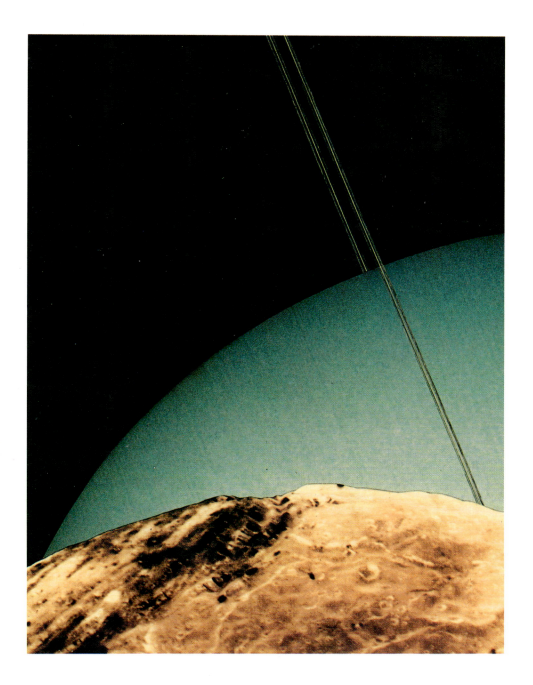

● In natural colour (far left), Uranus appears as a bland, pale blue ball in this Voyager 2 photograph. Computer-enhanced contrast and the addition of false colour (left) brings out the polar hood at the planet's south pole, here facing the camera.

One puzzling characteristic of Uranus is the exaggerated tilt of its axis, which lies almost in the plane of its orbit around the Sun. This means that for half the time during its 84 year orbit one or other of the poles is continuously in sunlight, while the opposite pole is in darkness. Presumably Uranus suffered some gargantuan collision early in its history that knocked it on its side. Whatever happened must have taken place before the planet's moons formed, because they orbit Uranus's equator, almost at right angles to the plane of its orbit.

So too do the planet's rings, which were discovered by accident in 1977 when astronomers watched Uranus pass in front of a star. Both before and after it disappeared behind Uranus the star blinked on and off several times as the rings, too faint to be seen directly from Earth, interposed themselves between us and the star.

For all the efforts of ground-based astronomers, Uranus remained largely unknown until the venerable Voyager 2 probe reached the planet in January 1986. The planet itself was disappointing – even from close range, there is little to see in the clouds of Uranus. Unlike Jupiter and Saturn, Uranus has only a weak internal heat source, so there is little convection to stir up cloud activity. According to one theory, Uranus has a large rocky core covered by a dense atmosphere of water vapour and ammonia topped by clouds of methane.

Voyager's photographs showed only the hemisphere of Uranus that was tilted towards the Sun at the time. Voyager scientists referred to this as the 'south' pole, because it lies below the plane of the planet's orbit, but the directions 'north' and 'south' mean little when dealing with such an extreme axial tilt. There is hardly any difference in temperature between this pole and the equator, so heat must be distributed effectively around the planet.

Over the planet's Sun-facing pole lay a dark haze produced by the action of sunlight on atmospheric methane, but this cleared towards the equator, revealing faint features in the clouds below. By tracking these clouds, astronomers were able to come up with the first reliable figures for the rotation of Uranus. At latitude 40°S the planet's clouds are carried around every 16 hours – over an hour faster than the rotation of the solid planet below, 17¼ hours, a figure determined by observations of the planet's magnetic field. Surprisingly, the rotation rate of the clouds gets slower towards the equator, not faster as on Jupiter and Saturn. At the equator of Uranus the clouds probably spin more slowly than does the planet's solid interior.

As if the extravagantly canted rotation axis were not enough, Uranus holds another surprise – its magnetic axis is tilted by 60° to the rotation axis. What's more, the magnetic axis is significantly off-centre, lying about one-third of the way from the planet's centre to its surface; why this should be so remains a mystery.

Inevitably, most interest focused on the rings and moons of Uranus. Voyager's discoveries brought the total number of known rings to eleven; to the five moons known previously Voyager added another ten, all smaller and closer to the planet. The two innermost moons orbit either side of the outermost ring, the Epsilon Ring, acting as shepherds. This ring varies in width from 20 to nearly 100 km (12 to 60 miles), but the narrowest rings are only 1 km (0.6 miles) wide. No one knows what keeps them so well-defined. Perhaps they are also flanked by shepherd moons too small and faint for Voyager to see.

Backlit by sunlight, fine dust was seen by Voyager to be thinly spread between the rings. This fine dust could not remain in orbit for long without settling into the atmosphere of Uranus, so it must be constantly generated by the disruption of larger bodies in the rings, from the impact of meteorites, and in mutal collisions. Like the rings, the

● Right: *Miranda, the smallest and innermost of the four major moons of Uranus, has strange grooved areas on its surface. Miranda may have been broken apart and reassembled in the past, possibly more than once.*

● Below left: *False-colour image of the rings of Uranus, showing nine rings (the pastel-coloured shading between them is due to computer processing). At the top is the thickest and brightest ring, the Epsilon Ring, about 100 km (60 miles) across at its widest.*

● Below: *A jagged valley on Miranda, the walls of which tower 15 km (10 miles) high, caused by faulting of the moon's crust. The largest impact crater in this picture is 25 km (15 miles) across.*

smallest moons are charcoal black, reflecting only a few per cent of the Sun's light, in stark contrast to the brilliant rings of Saturn. The darkness of their surfaces could be caused by the presence of carbon, or by methane that has been bombarded by high-energy protons in the magnetic field of Uranus, turning it into dark tar.

The four largest moons of Uranus are icy, cratered worlds, but the most astounding scenery was found on Miranda, at barely 500 km (300 miles) across the smallest of the moons visible from Earth. Its patchy surface consists of old, heavily cratered terrain interspersed with younger areas that are marked by strange grooves, resembling rake-marks. Perhaps the most striking feature is a fault valley 15 km (10 miles) deep, giving the moon's limb a jagged profile. Miranda looks as though it has been put together from different handfuls of modelling clay – and the startling conclusion is that the little moon was broken apart several times by major impacts early in its history, each time reassembling itself, only to be disrupted by further collisions.

Some of the other moons may have a similar history. The impacting objects may have been cometary nuclei, believed to have been more numerous in that part of the Solar System shortly after the formation of the planets. According to this theory, the smallest moons of Uranus, and the rings, would be fragments from the break-up of larger satellites. It seems that Uranus is a planet with a violent past.

"Formed a design, in the beginning of this week, of investigating, as soon as possible after taking my degree, the irregularities of the motion of Uranus . . . in order to find whether they may be attributed to the action of an undiscovered planet beyond it . . ." The writer was John Couch Adams, a Cambridge undergraduate, and the date was July 3, 1841. The "irregularities of the motion of Uranus" that he referred to were persistent deviations of the planet from the path that had been calculated for it. Astronomers had begun to suspect that Uranus was being drawn out of place by the gravitational attraction of an unknown planet, but Adams was the first to attempt to calculate its whereabouts.

He began by assuming the new planet lay at the distance specified by Bode's law (see p. 79). By the autumn of 1845 Adams had completed his calculations and sent them to the Astronomer Royal, Sir George Airy, at Greenwich. Unfortunately Airy was not convinced and did not order a search. Adams, a theorist who had no telescope, did not bother to look himself.

Meanwhile in France, Urbain Le Verrier had begun a similar investigation and sent the results to Airy in June 1846. His predicted position was only one degree away from that of Adams, whose work he did not know about. This time, though, Airy took notice. The Royal Observatory was too busy to undertake a search so Airy asked the astronomers at Cambridge to look for the new planet, but they made slow progress.

Becoming impatient, and receiving no help from astronomers in France, Le Verrier wrote to enlist the help of Johann Galle, a young and enthusiastic astronomer at Berlin Observatory. That same night, September 23, 1846, Galle and a student, Heinrich D'Arrest, set to work to check the sky around Le Verrier's predicted position. The German astronomers had the advantage of a brand-new star map. In less than an hour, they had found an object that was not on the chart.

The following night they confirmed the discovery. The object had moved, and it displayed a disk of just the size predicted by Le Verrier. "The planet whose position you have pointed out actually exists," Bode triumphantly wrote to Le Verrier on September 25. English astronomers were mortified. It is rare for astronomy to cause a national scandal, but it happened in England over the failure to discover Neptune.

From Earth, distant Neptune shows a small, greenish disk and little else other than two moons. Virtually everything we know about this world was taught to us in the summer of 1989 by the veteran space probe Voyager 2. Neptune is similar in size to Uranus and also has a blue–green colour caused by atmospheric methane. In another similarity, Neptune's magnetic axis is tilted at 47° to the rotation axis, and the centre of the magnetic field is offset from the planet's core. Evidently the magnetic field is generated in a layer outside the core. One major difference between these two planets is that Neptune gives out much more internal heat, which makes it more like Jupiter and Saturn; in this respect, Uranus is the odd one out among the four outer giants. Scientists are still a long way from understanding what goes on inside these planets.

Because of Neptune's stronger internal heat source, its clouds are far more active than those of Uranus. Voyager saw in the southern hemisphere a dark elliptical spot the size of Earth, the equivalent of Jupiter's Great Red Spot. The dark spot was fringed with white high-altitude clouds of methane cirrus, and completed its circuit of the planet in 18.3 hours. South of it lay a smaller dark spot with a white centre, presumably caused by rising and condensing gas. Between the two dark spots is a white cirrus cloud, dubbed the Scooter by Voyager scientists, which sped around the planet every 16¾ hours; despite its name, it actually rotated more slowly than the second dark spot, which moved around the planet in 16 hours.

● The three main characters in the discovery of Neptune. From top to bottom: John Couch Adams and Urbain Le Verrier, who independently predicted the existence of the planet, and Johann Galle, who actually discovered it from Berlin Observatory.

● *Neptune viewed by Voyager 2, showing the Great Dark Spot flanked by white cirrus clouds of methane.*

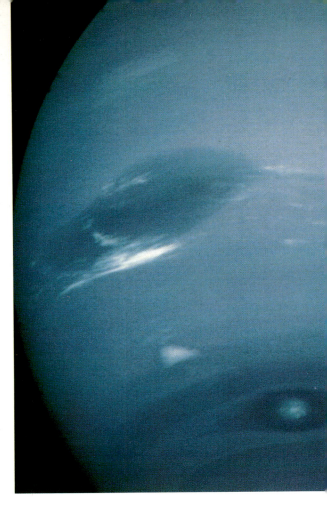

The composition of the main cloud deck is not certain, but it is thought to contain ammonia as well as hydrogen sulphide. Beneath the clouds, the solid body of the planet rotates in 16.1 hours, as revealed by observations of Neptune's magnetic field.

Scientists expected Voyager to find rings and more moons around Neptune, and they were not disappointed. Two narrow rings of dust loop around the planet 29,000 and 39,000 km (18,000 and 24,000 miles) above the cloud tops. Between them is a sheet of faint dust and another faint, broad ring further in. Voyager added six new moons to the two already known, but it was the close-up views of Triton, the largest of Neptune's moons, that stole the show, for it turned out to have active geysers of nitrogen.

With a diameter of 2700 km (1700 miles) Triton is larger than the planet Pluto, and was probably once an independent body that was captured by Neptune. Its strangely sculpted surface is covered with frozen nitrogen and methane at a temperature of −235°C, some of which evaporates to form a tenuous atmosphere. Atomic particles striking the surface turn the methane ice pink at the south polar cap, which is surrounded by a bluer fringe of fresh ice. Evidently the surface periodically melts and re-freezes with the changing seasons on Triton, smoothing out any impact craters.

● Above left: *High-level bands of cirrus cloud near the terminator of Neptune cast shadows onto the main cloud deck 50 km (30 miles) below.*

● Above: *The Great Dark Spot, an anticyclone the size of Earth. South of it is another dark spot with a bright core. Between the two dark spots is a bright cloud called the Scooter.*

● Left: *Neptune's rings, backlit by the Sun, are seen against a starry background in these two long exposures. The glare at the centre is caused by the overexposed image of Neptune itself. In addition to the two main rings a dusky inner ring can be seen, as well as a thin sheet of dust between the brighter rings.*

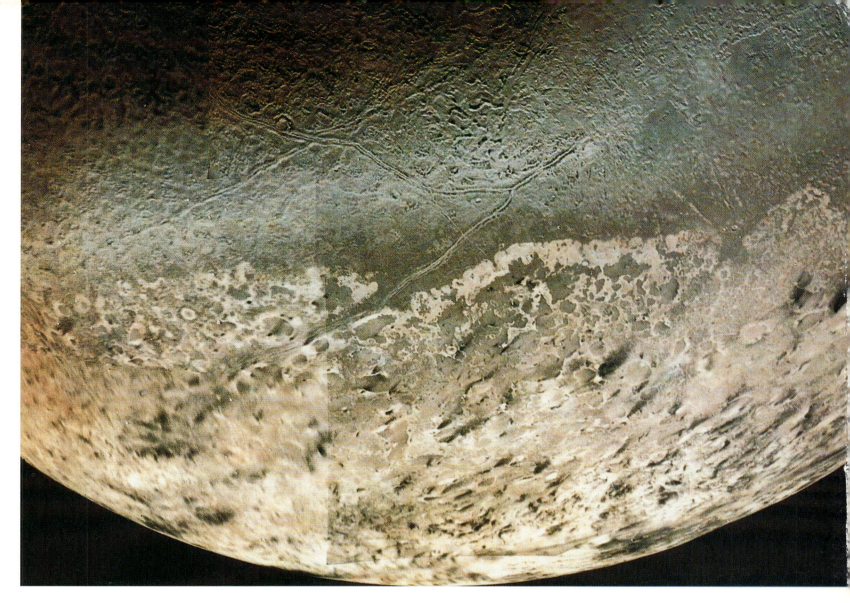

● Above: *Neptune's largest moon, Triton, has a pink polar cap, possibly of nitrogen and methane ice.*
● Below: *A depression on Triton, flooded by liquid that has since frozen.*
● Below right: *A dark streak from a nitrogen geyser on Triton (more are visible in the picture above).*

Most astoundingly of all, Voyager spotted the vents of dozens of ice volcanoes on Triton, some of which were caught in the process of erupting. These geysers are thought to be fed by underground pools of liquid nitrogen that erupt explosively through cracks, spraying plumes of material 8 km (5 miles) high. This material is blown downwind and deposited on the surface as dark streaks that are clearly visible in the Voyager pictures.

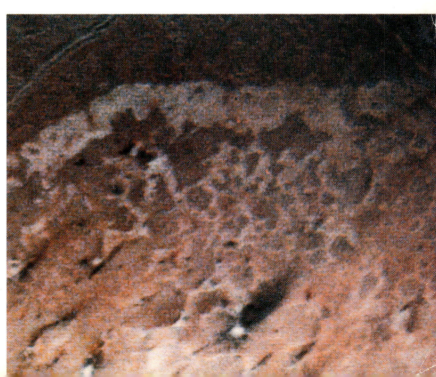

After the discovery of Neptune, astronomers experienced a sense of *déjà vu*. Like Uranus before it, the new planet seemed not to be following quite the path predicted for it, and several astronomers set out on the trail of yet another unseen planet. Foremost among them was Percival Lowell, best known for his observations of canals on Mars. Lowell had the advantage of owning his own observatory in Arizona, where he began his pursuit of Planet X, as he called it, in 1905.

This time, there could be no simple visual comparison with a chart. Planet X would be so remote, and hence so faint, that it would be impossible to distinguish it from the thousands of background stars by eye alone. Fortunately, photography came to the rescue. If the same area of sky were photographed twice, with a suitable time interval in between, a planet would give itself away by its movement between the two exposures. That, at least, was Lowell's hope, but many years of searching and calculation drew a blank. Lowell died in 1916, Planet X still undiscovered.

However, interest in Planet X did not die with him. The new director of Lowell Observatory, Vesto M. Slipher, decided to mount a fresh onslaught. Slipher had a new telescope built for the job, capable of photographing a wide angle of sky, and hired as an assistant an amateur astronomer with a keen interest in the planets, Clyde W. Tombaugh. It was Tombaugh's job to photograph the sky on large glass plates at night, and during the day to examine pairs of plates in a device called a blink comparator, which switched the operator's view rapidly back and forth from one to the other so that an object that had moved between exposures would seem to jump to and fro.

Tombaugh began the laborious and tedious work in April 1929. By the end of January the following year he had reached the constellation Gemini. On February 18, while blinking his way through a pair of photographic plates, Tombaugh recognized the tell-tale jump of a distant planet. He checked it on a third plate, taken to guard against photographic defects. There was no doubt – Planet X had been found.

● *Clyde Tombaugh, the American astronomer who discovered the planet Pluto in 1930. He is shown at the guiding eyepiece of the Lowell Observatory's 33 cm (13 inch) telescope with which the discovery plates shown on the opposite page were taken.*

● *Here lies a new planet: the starlike image of distant Pluto (arrowed) has moved between January 23, 1930 (left) and January 29. These are small sections of the plates with which Clyde Tombaugh made the discovery on February 18, 1930.*

That night was cloudy. But over the following nights the area was re-photographed, and the object was also examined visually. The discovery was announced on March 13, 1930, the 75th anniversary of Lowell's birth, and 149 years to the day since Herschel had discovered Uranus. The new planet was named Pluto, after the god of the underworld, appropriate for the darkness in which it orbits at the edge of the Solar System.

There was, however, a nagging problem. Pluto was much fainter than Lowell or anyone else had expected. Worse still, it did not show a disk under even the highest magnification; clearly, it was much smaller than Uranus or Neptune. Exactly how small was not fully realized until nearly half a century later. Astronomers made two discoveries in the late 1970s that significantly changed their impressions of Pluto. Firstly, methane frost was detected on its surface, indicating that it was more reflective than previously imagined. Secondly, and most important of all, a moon of Pluto was discovered in 1978.

This moon was named Charon, after the mythological boatman who ferried dead souls into the underworld. Mutual eclipses of Pluto and Charon have allowed astronomers to assess their diameters: 2200 km (1350 miles) for Pluto and 1200 km (750 miles) for Charon. Charon, then, is fully half the diameter of Pluto, making it the largest moon in the Solar System relative to the size of its parent planet. Pluto's diameter is only two-thirds that of our own Moon, and its mass is less than one-fifth the Moon's. Not only is Pluto by far the smallest planet, but some astronomers have even questioned its right to planetary status at all, suggesting that it is simply an icy leftover from the formation of the outer planets. Perhaps it is simply the largest of an unseen family of asteroids in the outer reaches of the Solar System.

Another peculiarity of Pluto is its eccentrically shaped orbit, which for 20 out of every 248 years brings it within the orbit of Neptune. Pluto crossed Neptune's orbit in 1979 and reached its closest point to the Sun, perihelion, in September 1989. It will not have receded beyond Neptune's orbit again until 1999, so Neptune is temporarily the outer-most planet. Pluto is the only planet whose orbit is this wayward. There is no danger of Neptune and Pluto colliding, since their orbits are inclined like two separate hoops and so do not actually intersect.

Pluto spins slowly, every 6.4 days, and Charon orbits it in the same time, permanently fixed a mere 20,000 km (12,000 miles) above one hemisphere like a geostationary satellite. According to one theory, Charon is a chunk that was knocked off Pluto in an ancient collision. There are no plans to send a space probe to Pluto, but probably the planet is similar to Neptune's largest moon, Triton.

● Left: *Pluto and its moon Charon lie far from the Sun, shown as a bright yellow star at top right in this artist's impression. Pluto's surface is thought to be covered with nitrogen and methane ice, as is Neptune's moon Triton, sculpted into weird shapes by cycles of thawing and freezing as Pluto moves along its elliptical orbit. When Pluto is closest to the Sun, the surface warms up enough for some of the ice to evaporate and produce a thin atmosphere.*

Somewhere in the darkness beyond the outermost planets orbits an immense swarm of objects with the composition of 'dirty snowballs', a few hundred metres to several kilometres in diameter. Occasionally, the gravitational wash of a passing star nudges some of these bodies out of the swarm and they begin to fall in towards the Sun. As the snowball approaches the Sun it warms up, releasing gas and dust to form an enveloping halo that can be seen from Earth as a comet.

Although the popular image of a comet is of a glowing body with a long tail, many comets are tailless, and the great majority of them never become bright enough to be seen without a telescope. Every year some two dozen comets are recorded, about half of them new discoveries, and the rest previously known comets making a reappearance. Many new comets are found by amateur astronomers. In so doing the astronomers achieve immortality, for a comet is named after its discoverer, or up to three co-discoverers for multiple sightings, which accounts for such tongue-twisters as Comet Sugano–Saigusa–Fujikawa and Comet Honda–Mrkos–Pajdusáková.

Left to its own devices, a comet would simply swing around the Sun and recede into the depths of space on a highly elongated orbit, not to return again for millions of years. However, some comets are diverted by the gravitation of the planets, notably Jupiter, into shorter orbits so that they return every 200 years or less; these are termed *periodic comets*. The most famous example is Halley's Comet, which orbits the Sun every 75 years. Encke's Comet has the shortest known orbital period, at 3.3 years.

At the heart of a comet is its *nucleus*, the 'dirty snowball' that releases gas and dust to form an enveloping *coma*. A comet's coma is the size of a planet but far thinner than the Earth's atmosphere – so thin, in fact, that stars can be seen shining through it. Together, the coma and nucleus make up the comet's head.

● Below: *Two types of comet tail were well displayed by Comet West, one of the most prominent comets of the century, in March 1976. The broad dust tail appears white and the narrower, straighter gas tail is blue.*

● Left: *Jets of gas and dust squirt from the 'dirty snowball' nucleus of Halley's Comet, 15 km (10 miles) long, photographed by the Giotto probe in March 1986.*

● *The Star of Bethlehem was depicted as a comet (the reddish object at top centre) in* The Adoration of the Magi *by the Italian artist Giotto di Bondone. He may have modelled it on Halley's Comet which appeared in 1301, shortly before he painted this fresco. The European space probe to Halley's Comet in 1986 was named Giotto after him.*

The European Space Agency's probe Giotto gave us our first close-up look at a comet's nucleus when it flew through the head of Halley's Comet in March 1986. Giotto's photographs showed that the nucleus was shaped like a lumpy potato, 15 km (9 miles) long and 8 km (5 miles) wide, with jets of gas and dust spraying out through cracks in the dark crust. Beneath its dusty crust the nucleus consisted mostly of frozen water, and at the time of the Giotto encounter it was giving off enough water vapour to fill a small swimming pool every second.

Large and bright comets such as Halley's have two types of tail, one composed of gas and the other of dust; the relative prominence of the two types of tail varies from comet to comet. The gas tail is blown away from the coma by the solar wind, the stream of atomic particles from the Sun, whereas the dust tail is caused by the pressure of sunlight on dust particles. Both tails always point away from the Sun, no matter which direction the comet is moving in. The gas tail is usually straight, narrow and bluish, but dust tails can spread out into a spectacular curve or fan shape.

IS THERE A TENTH PLANET?

After the discovery of Pluto, Clyde Tombaugh continued his photographic survey for additional planets until he had covered a broad band right the way round the sky. If any other large planets existed out there, he would almost certainly have found them. Clearly, Pluto is not large enough to have affected the motions of Uranus or Neptune, so no calculations could have predicted its existence. The discovery of Pluto was an accident, due simply to the thoroughness of Tombaugh's search. From time to time astronomers have predicted the existence of a tenth planet, but nothing has been found near the calculated positions. It seems most likely that the supposed irregularities in the orbits of Uranus and Neptune are not real at all but simply the results of observational errors. If there is a tenth planet it has hidden itself very well.

5 EARTH AND MOON

Turn a pair of binoculars on the Moon and run your eye along the terminator, the line of demarcation between lunar day and night. Across the 384,000 km (240,000 miles) of space that separates us from our nearest natural neighbour, the main features near the terminator stand out clearly – mountains, craters, and valleys – thrown into exaggerated relief by the oblique sunlight. There is no atmosphere to soften the harsh black shadows on this barren, rocky ball, 3476 km (2160 miles) in diameter, a quarter the size of the Earth.

Look again the following night and the terminator will have moved, bringing new features into view. As the Sun climbs higher into the lunar sky the shadows shorten, lessening the impression of roughness. When the Moon is full, and the Sun illuminates its surface from directly overhead, the terrain seems to have been rolled flat. What we then see is not so much the surface topography, but differences in contrast between different types of surface material.

The darkest areas are the lowland plains, known as *maria* (singular, *mare*), Latin for seas, since that is what they were once thought to be; now we know that they are ancient lava-flows ground down into dust by aeons of meteorite bombardment. The maria are arranged in the familiar pattern visible to the naked eye as the 'man in the Moon'. Although they are the smoothest areas, even they are not unblemished by the Moon's most ubiquitous feature: craters, formed by impacting meteorites and comets.

Craters of all sizes abound on the Moon, the largest of them 200 km (125 miles) or more across. Most craters are named after great scientists of the past, whereas the names of the maria are unlikely sounding abstractions, such as Mare Tranquillitatis (Sea of Tranquillity), Oceanus Procellarum (Ocean of Storms), and Mare Imbrium (Sea of Rains). There is no real distinction between a large crater and a small mare such as Mare Crisium (Sea of Crises), 500 km (300 miles) across. Without an atmosphere to help protect it, the Moon is wide open to bombardment from space. And, since there is neither wind nor water, the agents of erosion that have shaped the Earth's surface, features such as the footprints of the Apollo astronauts will remain unchanged for hundreds of millions of years.

● *The Moon is the Earth's nearest natural neighbour in space. Its changing phases are a familiar feature of the night sky, and its gravity raises tides in the oceans.*

The oldest parts of the surface, the bright uplands, bear witness to the heavy battering the Moon suffered in its youth. Many craters are overlain by smaller ones caused by later impacts. Large craters often have complex, terraced walls and central mountain peaks. Some, around the edges of the maria, have been flooded with dark lava and their walls eroded. Were we to stand inside a large crater we would not be able to see its walls, for the curvature of the Moon would take them below the horizon.

Bright rays of ejected material splashed out by impacts extend for many hundreds of kilometres from certain craters, most notably Tycho in the southern hemisphere and Copernicus near the equator. As lunar features go, these craters are young – about 850 million years in the case of Copernicus, and perhaps only 100 million years for Tycho. Many other craters threw out pulverized rock when they formed, but the rays have subsequently faded.

The main features visible with binoculars and small telescopes are shown on the maps on pages 116–119. These maps show the Moon the 'right' way up, as it appears to the naked eye or through binoculars. To match the view through a telescope, which gives an inverted image, turn the maps upside down.

● *A view of the Moon not obtainable from Earth. This photograph was taken by the Apollo 17 astronauts on their return flight to Earth, and shows part of the far side, which is permanently averted from our sight.*

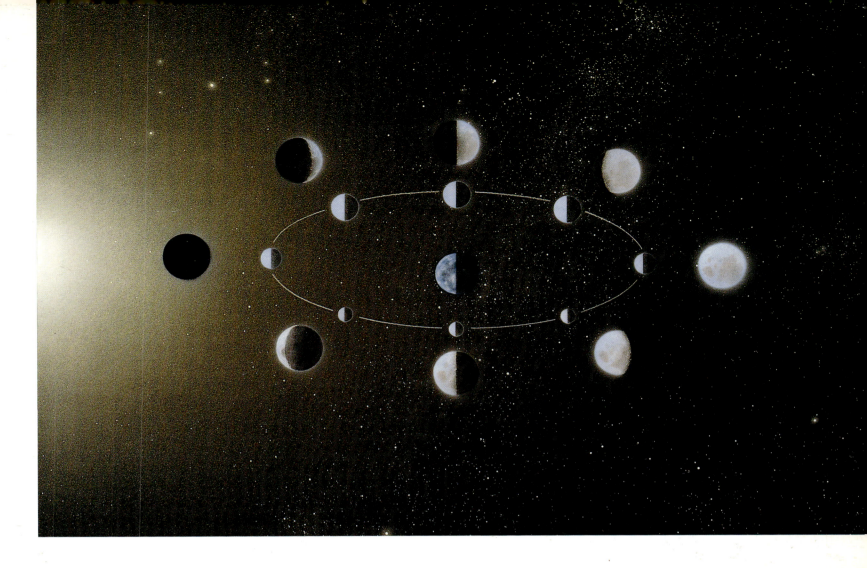

● *The Moon goes through a cycle of phases each month. As it orbits the Earth we see different amounts of its sunlit hemisphere, from new (invisible) through crescent, half, gibbous, full, and back to new again.*

The Moon is reined into permanent orbit by the Earth's gravity. From night to night the Moon goes through its cycle of phases as it orbits the Earth. The sequence begins and ends at new Moon, when it lies between the Earth and the Sun. First it emerges into the evening twilight as a slender crescent. Sometimes the night-time portion of the crescent Moon can be seen faintly illuminated by light reflected from the Earth, giving rise to the appearance popularly known as the 'old Moon in the new Moon's arms'.

Over the next few nights the crescent fills out until the Moon is half illuminated, which is known as first quarter since the Moon is a quarter of the way around its orbit. The terminator then becomes convex as the Moon passes through its gibbous phase to become full, nearly 15 days after new Moon; it then lies on the opposite side of the sky to the Sun. After full Moon the phases repeat in reverse order: gibbous, third quarter, then a shrinking crescent that retreats ever closer to the Sun in the morning sky, finally reaching new Moon again. The whole cycle takes 29½ days to complete and is the origin of our calendar month.

Glances at the Moon from one month to another will confirm that the same features are always visible as it orbits the Earth – the Moon's axial spin and orbital movement are synchronized so that it keeps one side permanently facing us. This is no coincidence. The Earth's gravity has braked the Moon's spin to produce what is known as *captured rotation*, and the same thing has happened to the moons of many other planets. As a result, any given place on the Moon experiences two weeks of daylight, during which the surface temperature reaches 110°C, followed by a two-week night when the temperature drops to −170°C.

Because of this captured rotation, it was not until space probes were sent around the Moon that we obtained our first views of the mysterious far side, forever hidden from us here on Earth. It turned out to be much more rugged than the near side, consisting almost entirely of bright, cratered uplands with few dark maria. Evidently the crust on the Moon's far side is a few kilometres thicker than on the near side, which made it more difficult for lava to flood out from the lunar interior.

The Moon's gravity affects the Earth, most notably in the rhythmic rise and fall of the ocean tides. Both the Sun and the Moon raise tides, but the Moon's effect is over twice as great because it is closer to us. When the Moon and Sun pull in line, tides are highest; these are termed *spring tides*, and occur at full Moon and new Moon. But when Moon and Sun pull at right angles, at first quarter and last quarter, the tides are less pronounced; these are known as *neap tides*.

Two bulges occur in the Earth's oceans, one on the side facing the Moon where its gravitational attraction is greatest, and one on the opposite side where the Moon's attraction is least. Most places have two tides a day as the Earth rotates under these bulges. Tides occur approximately 50 minutes later each day because of the daily movement of the Moon in its orbit.

Local coastlines have a considerable effect on the height of tides. For example, the Atlantic waters are funnelled into the Bay of Fundy in Canada, producing a range of as much as 15 metres (50 feet) between high and low tide, but this is extreme. The importance of tides should not be overlooked. Global shipping gears its activities to the tempo of the tides. Many forms of life depend on the sweep of the tides up and down the shoreline. Indeed, life may owe its very existence to the ebb and flow of ancient oceans that repeatedly washed a rich soup of chemicals over the Earth's primeval beaches where they combined to form the first complex molecules of life, about four billion years ago.

The tides have another, more subtle effect – they are slowing down the rotation of the Earth, so that the day is gradually getting longer. The amount is very small, less than two thousandths of a second per century, but it adds up: in a little over 200 million years from now there will be 25 hours in a day.

● *The Earth's ocean tides rise and fall in response to the Moon's gravitational pull. Most places on Earth have two tides a day and the tides recur approximately 50 minutes later each day.*

● How eclipses occur. When the
Moon's shadow falls on the Earth
(left), a solar eclipse occurs; when the
Moon enters the Earth's shadow
(right), there is a lunar eclipse.

Occasionally, the Moon moves in front of the Sun at new Moon, or
passes into the shadow of the Earth at full Moon, thereby producing an
eclipse. The reason why this does not happen at every new and full
Moon is that the Moon's orbit is tilted with respect to the Earth's; only
where the orbits intersect can the Sun, Moon, and Earth line up exactly.

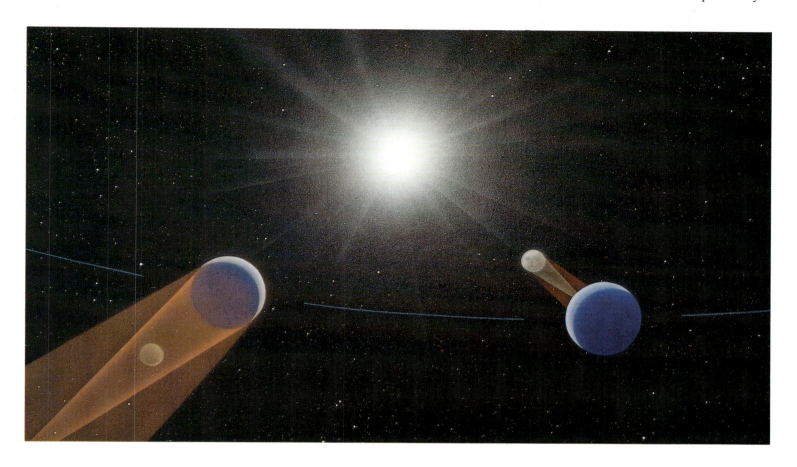

A total eclipse of the Sun is one of the most spectacular phenomena
in nature, so rare and beautiful that astronomers will travel across the
world to view one. It begins as the Moon glides across the Sun's face,
taking a 'bite' out of the Sun. Indeed, the ancient Chinese believed that
eclipses were caused by a dragon attempting to devour the Sun, and they
banged gongs to frighten the creature away, a technique that never
failed.

Almost imperceptibly, the eclipse progresses for over an hour until
the sky becomes noticeably dark. Birds and other animals, fooled by the
deepening twilight, bed down for the night. Finally, only a thin crescent
of sunlight is left, which for the last few seconds is broken up into a
necklace of brilliant points by mountain peaks along the Moon's edge.
This effect is known as *Baily's beads* after the English astronomer
Francis Baily, who described it at an eclipse in 1836. Often one bead
shines more brightly than the others to give what is called the *diamond
ring effect*. Then the Moon completely masks the Sun and the eclipse
becomes total, bright stars and planets appearing in the darkened sky.

A total eclipse brings into view the Sun's corona, a faint halo of gases
whose pearly light is normally swamped by the much greater brilliance
of the Sun's disk. In addition, a few pinkish prominences may be seen
like tiny blossoms around the Sun's rim. Totality can last anything from
a few seconds up to seven-and-a-half minutes, although the usual
duration is just a few minutes. Then the flash of the diamond ring and
Baily's beads reappear, announcing that the eclipse is over, and the
Moon moves off the face of the Sun like a curtain slowly being drawn
back.

Total solar eclipses can be seen only from within the narrow band on which the dark central portion of the Moon's shadow falls, usually no more than about 200 km (125 miles) wide, so they are rare from any given place on Earth. However, partial eclipses – visible either side of the band of totality – may be seen from a much wider area. At least two eclipses of the Sun are visible somewhere on Earth each year, and there can be as many as five.

That the Moon and the Sun appear virtually the same size in the sky is a remarkable coincidence. But the Moon's apparent size changes somewhat, because its orbit is elliptical and so its distance from the Earth varies. When the Moon is at its farthest from Earth it is too small to cover the Sun completely. The result is an *annular* eclipse, so called because a ring, or annulus, of sunlight is left around the Moon's disk. These are of much less interest to astronomers because the solar corona does not become visible.

Whereas an eclipse of the Sun is possible only at new Moon, an eclipse of the Moon can happen only at full Moon. Lunar eclipses last longer than solar ones and they are visible over a much wider area. Up to three occur each year. They are of little scientific interest, but they provide an attractive and easily observed example of nature's clockwork in action.

● Left: *The brilliant diamond ring effect occurs when only a tiny part of the Sun's disk is unobscured, immediately before or after a total eclipse. This eclipse was photographed from Hyderabad, India, in February 1980.*

A lunar eclipse begins with a darkening at the rim of the full Moon as it moves into the Earth's fuzzy shadow projected into space. For the next hour or more the Moon moves deeper into the shadow until it is fully immersed. To an observer on the Moon, the Sun would appear to be totally eclipsed by the Earth.

Strangely enough, the Moon usually does not vanish from view when totally eclipsed because some light is refracted onto it through the Earth's atmosphere, giving the eclipsed Moon a dark red or coppery colour, as though steeped in tannin. Only rarely does the Moon become so dark that it vanishes at mid-eclipse. This happens when the Earth's atmosphere is exceptionally cloudy or if there is a lot of high-altitude dust from volcanic eruptions, blocking the passage of sunlight into the Earth's shadow. Totality can last an hour or more before the Moon starts to emerge from the shadow.

● Above: *Multiple exposure of a lunar eclipse, showing the entry of the Moon into the Earth's shadow from left to right and its subsequent exit. The Moon shines red even during the eclipse because sunlight it refracted onto it through the Earth's atmosphere. This eclipse took place in July 1982.*

● Left: *A partial eclipse of the Sun, photographed at sunset from Siding Spring Observatory, Australia, in April 1986. Partial eclipses can be seen from a much wider area on Earth than can total eclipses.*

THE EARTH'S ORBIT AND ITS SEASONS

The Earth orbits the Sun at an average distance of 149,600,000 km (92,960,000 miles), which astronomers term the *astronomical unit*. A year lasts just under 365.25 days, and the odd fraction is accounted for by adding a day to the calendar every leap year. The differing amounts of sunlight received by each hemisphere during the course of the year give rise to the seasons.

The Earth's axis is not upright with respect to its orbit but is tilted by 23½°, so the north and south poles alternately lean towards and away from the Sun as the Earth moves around the Sun. At the *summer solstice* the northern hemisphere is tilted at its maximum towards the Sun. This happens on or around June 21 (the exact date varies because of the effect of leap years). Six months later, at the *winter solstice* on about December 22, the tilt favours the southern hemisphere. In between are the *equinoxes* (roughly March 21 and September 23), when the Sun lies overhead at noon on the equator.

As the Earth's orbit is slightly elliptical its distance from the Sun varies, by 5 million km (3 million miles), over the year, but this is not the cause of the seasons. The Earth is closest to the Sun (at perihelion) in early January, during the northern hemisphere's winter, and at its most distant (aphelion) in early July, northern summer.

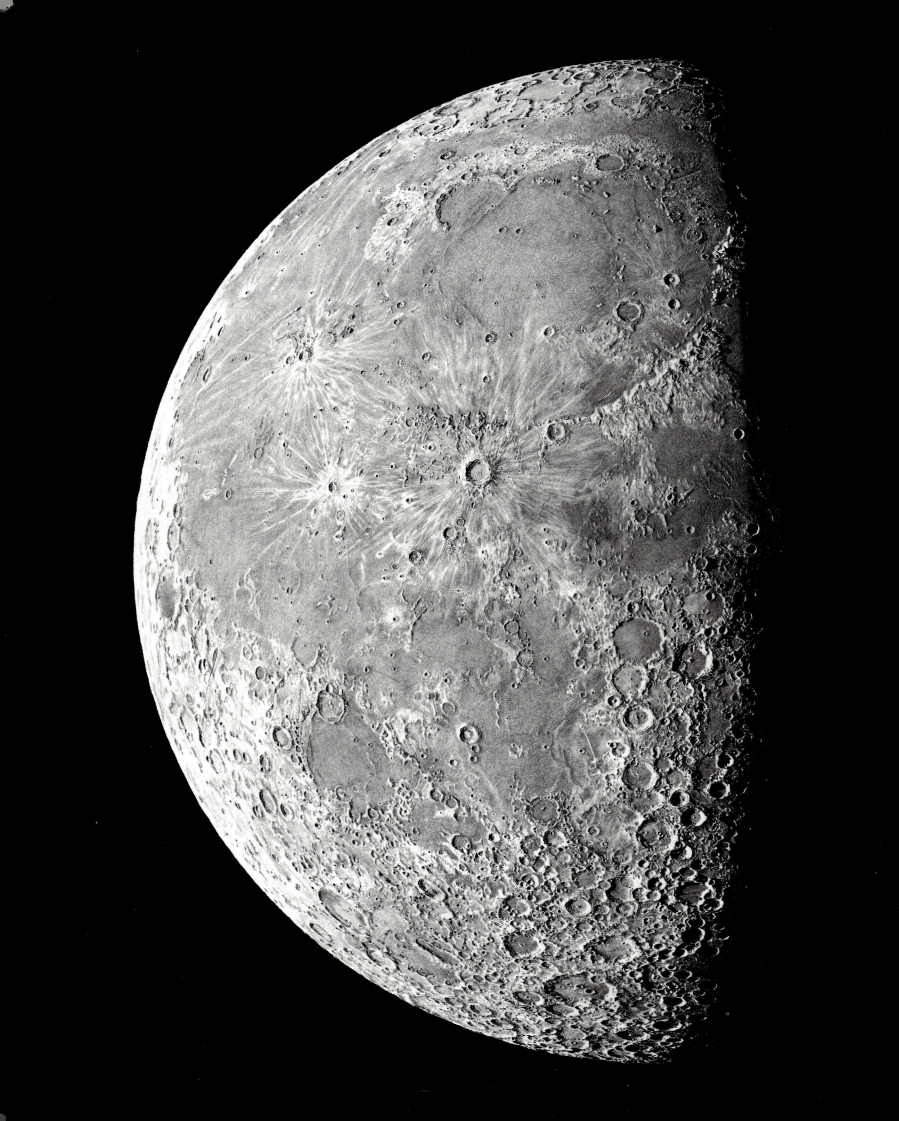

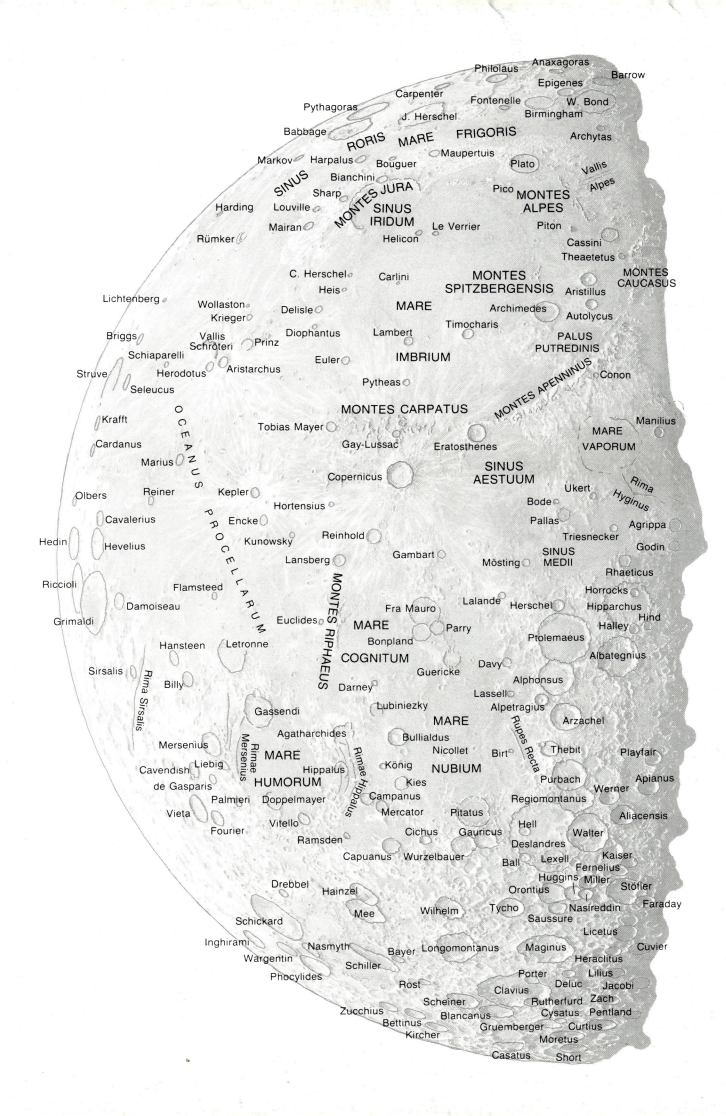

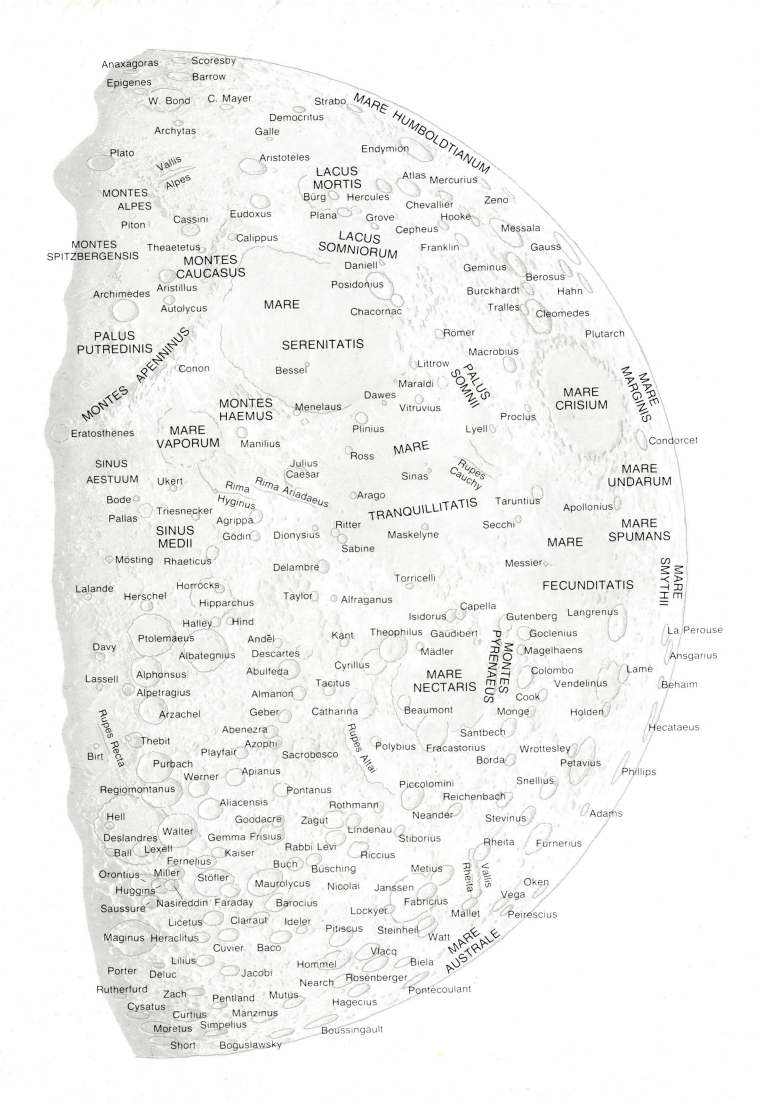

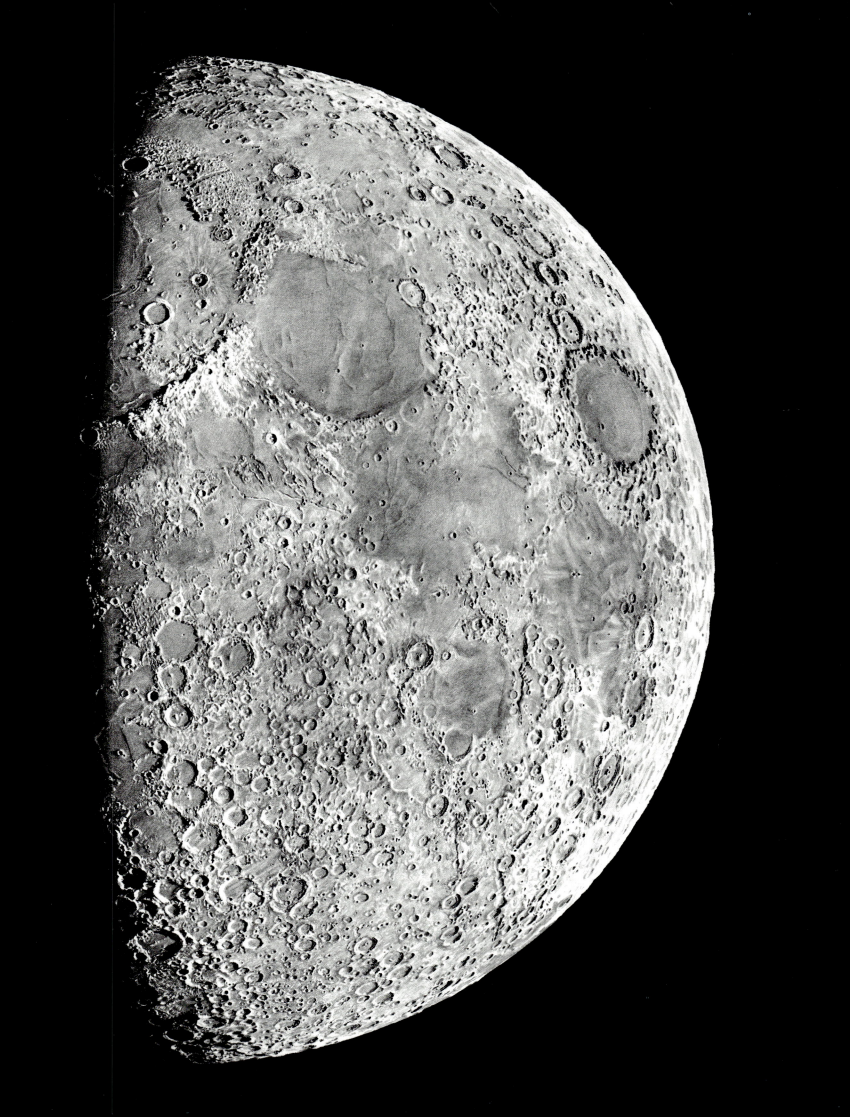

The Earth and Moon can be thought of as a double planet, for the Moon is unusually large – over a quarter the size of the Earth, a ratio exceeded only by Pluto and its moon Charon. All the other planets in the Solar System are many times larger than their moons. How did the Earth come to have such a disproportionately large satellite?

Before the Apollo astronauts brought back lunar rocks for study there were three main theories of the Moon's origin. Last century George Darwin, son of the the naturalist Charles Darwin, proposed that the Moon was once part of the Earth that was flung off long ago when the Earth was spinning rapidly, every few hours (see illustration). This is known as the *fission theory*. However, it is now thought highly unlikely that the Earth could ever have rotated quickly enough to have split up as proposed.

Fission theory

Moon

Earth

Earth's orbit

Earth

Impact

Capture

How the moon may have formed.
● Far left: *In the fission theory, the Moon split off from the hot, rapidly spinning young Earth.*
● Right: *Another theory says that the Moon built up from a ring of particles orbiting the Earth.*
● Below left: *According to the capture theory, the Moon was once a separate body that wandered too close and became trapped by the Earth's gravity. The leading theory at present says that the Moon was formed from debris thrown off by the impact of a Mars-sized body with the Earth.*

According to the *capture theory*, the Moon was once an independent body that strayed too close to the Earth and was trapped into orbit by the Earth's gravity. The tiny moons of Mars and the outer moons of Jupiter and Saturn are believed to have been snared in this fashion. But it seems almost impossible for such a large body as our Moon to have been captured by the Earth.

The most popular theory said that the Earth and Moon formed side by side, as they are today, with the Moon being built up from a disk of orbiting matter. This is how the Galilean moons of Jupiter and the main moons of Saturn and Uranus are thought to have originated. However, there are differences in composition between the materials that make up the Earth and the Moon that are difficult to explain by this theory.

In short, none of the three theories was convincing. Since the Apollo missions, support has grown for a fourth theory – the Moon formed from the debris of a monumental collision between the Earth and a body the size of Mars. This proposal builds on the growing realization that catastrophic impacts may have been common during the final stages of the formation of the planets. It was then, says the new theory, that the Earth was struck a glancing blow by a body about one-tenth its mass. The collision would have vaporized the impacting body and part of the Earth's outer layers, blasting large amounts of material from both of them into orbit around the Earth. A ring of material was created that subsequently formed into the Moon.

This idea combines the strongest aspects of the previous three, successfully accounting for the difference in composition between the Earth and Moon, while explaining how the Moon came to be in its present orbit. It seems to be the best explanation for the Moon's origin yet.

"Tranquillity Base here. The Eagle has landed." With these words Apollo 11 astronaut Neil Armstrong announced the success of the first manned lunar landing. The time was 8.18 pm GMT, on July 20, 1969. Thus ended the quest that had begun in 1961 with President John F. Kennedy's promise that the United States would put a man on the Moon and return him safely to Earth "before this decade is out."

The Apollo 11 adventure had begun four days earlier when Armstrong, Michael Collins, and Edwin Aldrin blasted off from NASA's spaceport at Cape Canaveral in Florida atop a Saturn V rocket. They were crammed into a conical capsule 3.9 metres (12.8 feet) across at its widest. Once they had left the Earth and were safely on their way to the Moon, the astronauts turned their command module to dock with the spidery lunar module, nicknamed Eagle, in which Armstrong and Aldrin would make the actual landing.

● Above: *Apollo 11 astronaut Edwin Aldrin on the Moon. Neil Armstrong and the lunar module Eagle are reflected in his visor.*

● Right: *Apollo 11 lifts off from Cape Canaveral atop a Saturn V rocket.*

● Top right: *Separation of the first stage of the Saturn V.*

● Far right: *The top half of the lunar module carrying Armstrong and Aldrin returns to orbit from the Moon's surface, the landing accomplished and the waiting Earth in the background.*

Three days after launch the astronauts went into orbit around the Moon and gazed down towards their intended landing site in Mare Tranquillitatis. The following day, Armstrong and Aldrin crawled through a hatch into the lunar module and separated it for the descent to the surface, leaving Michael Collins alone in the command module. While people around the world followed the event on TV and radio, Armstrong cautiously piloted Eagle down to a safe landing, taking over from the automatic landing system when it became clear that Eagle was headed for a crater filled with boulders. Less than thirty seconds' worth of fuel remained when they touched down.

● *Aldrin descends the ladder from the lunar module to the Moon's surface.*

● Above: *The crater Aristarchus, from orbit.*

Around them was a flat, rock-strewn landscape pitted with small craters. Armstrong, the mission commander, was the first to leave the lunar module. As he stepped down to the surface he spoke the now famous words, "That's one small step for man; one giant leap for mankind." (Actually, he fluffed his lines – he missed out the 'a' in 'for a man'.) Armstrong reported that the dark grey lunar soil stuck to his boot like powdered charcoal. Later studies confirmed that the soil lay up to 20 metres (60 feet) deep over the solid bedrock.

Aldrin joined him on the surface and for the next two hours they proceeded to set up experiments and collect 20 kg (48 lb) of rock and soil samples for scientists on Earth to analyse. Moving around on the Moon, where gravity is one-sixth that on Earth, was easy even though they were wearing bulky spacesuits. They unveiled a plaque on one of the legs of the lunar module which bore a picture of both the western and eastern hemispheres of the Earth, and the words "Here men from the planet Earth first set foot upon the Moon. We came in peace for all mankind."

Their tasks completed, the explorers returned to the lunar module for a meal and a somewhat cramped night's sleep. Less than 24 hours after touching down on the Moon, Armstrong and Aldrin took off in the upper half of the lunar module to rejoin Michael Collins orbiting above.

Jettisoning the rest of Eagle, the astronauts set course for Earth, and

● Below: *The Apollo 16 lunar rover and lunar module.*   ● *Apollo 15 astronaut James Irwin with the lunar rover at Mount Hadley.*

splashed down in the Pacific on July 24. Even so, their ordeal was not quite over. For the next two-and-a-half weeks they remained in quarantine while scientists tested the Moon samples to see if they contained any germs that might contaminate the Earth. Fortunately, the Moon rocks proved to be completely sterile and the quarantine procedure was abandoned for future Apollo crews.

In composition, the Apollo 11 samples turned out to be broadly similar to volcanic basalt on Earth, although with some distinctive chemical differences, one being that there is absolutely no water at all in the Moon rocks. Even the sands of the Sahara are wetter than the surface of the Moon. This was not too surprising. What was astounding was their extreme age: over 3000 million years, far older than all but the most ancient rocks on Earth. And Mare Tranquillitatis is one of the *youngest* areas on the Moon.

While scientists were coming to terms with these first findings, a second crew set out for the Moon. Apollo 12 landed in Oceanus Procellarum, a short walk away from an unmanned pathfinder probe, Surveyor 3, that had landed there two years earlier as part of the reconnaissance for the Apollo missions. Astronauts Charles Conrad and Alan Bean brought back parts of the long-dead robot probe for engineers on Earth to see how it had withstood its long exposure to the harsh conditions of space.

Travel to the Moon was starting to seem routine – but then disaster struck. On its way to the Moon, unlucky Apollo 13 suffered an explosion in an oxygen tank in the equipment section behind the conical command module. Unable to turn their spacecraft round, the astronauts had no option but to continue to the Moon and let its gravity swing them back to Earth. They were running desperately short of air, water, and electricity, but fortunately the lunar module was still attached. Its supplies saved them, and they made it safely home. Apollo flights were suspended while the spacecraft was redesigned to prevent a recurrence of the disaster.

With each successive mission, astronauts spent longer on the Moon. Apollo 15 carried the first lunar car, known as the lunar rover, which was capable of speeds of up to 14 kph (9 mph) and allowed a much wider area to be explored. It was powered by batteries and had wire wheels, so there was no chance of punctures. David Scott and James Irwin drove to the edge of Hadley Rille, a deep valley carved out by flowing lava at the foot of the Apennine Mountains.

Apollo 16 made the first landing in the highlands, while Apollo 17, the last of the series, set a host of records. It was the longest Apollo mission, lasting over 12 days from launch to splashdown. Eugene Cernan and Harrison Schmitt spent three days on the Moon, driving 35 km (22 miles) and collecting 110 kg (240 lb) of rocks. Apollo 17 returned to Earth on December 19, 1972. No humans have been to the Moon since.

In all, the six successful Apollo missions returned 380 kg (840 lb) of rock, much of which is still stored under sterile conditions awaiting analysis. Divided into the $40 billion cost of the whole Moon programme, this makes the rocks worth nearly $100 million per kilogram, many times their weight in gold. The rocks of the maria are rich in iron, titanium, and magnesium, which may make them suitable for mining, while the highland rocks contain more calcium and aluminium.

From detailed studies of the Apollo samples, scientists have established that the Moon is the same age as the Earth, 4600 million years (a difference of a few million years would be too small to measure). For the first 700 million years of its existence the Moon was subjected to a heavy bombardment by asteroids and meteorites that left

THE APOLLO MISSIONS TO THE MOON

Mission	Launch date	Astronauts	Remarks
Apollo 11	July 16, 1969	Neil Armstrong Michael Collins Edwin Aldrin	Armstrong and Aldrin made first manned lunar landing in Mare Tranquillitatis, July 20
Apollo 12	November 14, 1969	Charles Conrad Richard Gordon Alan Bean	Conrad and Bean landed in Oceanus Procellarum, November 19
Apollo 13	April 11, 1970	James Lovell John Swigert Fred Haise	Lunar landing cancelled after explosion on board spacecraft
Apollo 14	January 31, 1971	Alan Shepard Stuart Roosa Edgar Mitchell	Shepard and Mitchell landed near crater Fra Mauro, February 5
Apollo 15	July 26, 1971	David Scott Alfred Worden James Irwin	Scott and Irwin landed on July 30 near Hadley Rille; first use of lunar rover
Apollo 16	April 16, 1972	John Young Thomas Mattingly Charles Duke	Young and Duke landed in highlands near crater Descartes on April 21
Apollo 17	December 7, 1972	Eugene Cernan Ronald Evans Harrison Schmitt	Cernan and Schmitt landed on December 11 at edge of Mare Serenitatis

● Far left: *Apollo 14 astronaut Edgar Mitchell sets up experiments on the Moon.*

● Top left: *Harrison Schmitt collects samples with a scoop on the Apollo 17 mission.*

● Left: *Moon rocks brought back by Apollo 11.*

● Above: *The Apollo 16 lunar module blasts off from the Moon.*

● Below: *Splashdown. Frogmen attach a flotation collar to the Apollo 14 command module in the Pacific Ocean.*

its crust scarred and cratered. This bombardment ended in a crescendo about 3900 million years ago when giant impacts dug out the major mare basins.

Heat released by radioactive elements began to melt the interior of the Moon. Dark lava then began to seep out, filling the mare basins. This process continued until about 3000 million years ago, which is the youngest age of any of the Apollo samples. Since then, the Moon has remained largely unchanged apart from the arrival of the occasional meteorite large enough to produce a crater visible from Earth.

However, localized lava-flows may have continued to erupt until about 1000 million years ago, judging by the youthful appearance of parts of the surface. Even now, the Moon may not be entirely dead, for observers have reported occasional obscurations or brightenings that may signal the release of gases through cracks in the crust.

Next century humans will return to the Moon to set up bases similar to scientific stations in Antarctica. These bases will probably be buried under layers of lunar soil to insulate them from dangerous solar radiation and cosmic rays. Their inhabitants will breathe oxygen extracted from the lunar soil, where it is locked up chemically. Scientists will come to the bases to learn more about the Moon itself, to observe the Universe through telescopes, and to carry out experiments in the vacuum and low-gravity conditions. Eventually the Moon will be used as a staging post for expeditions to other parts of the Solar System. One day, tourists will take exotic holidays on the shores of the Sea of Tranquillity, taking in a visit to the historic Apollo 11 landing site where the footprints of the first astronauts will be preserved as part of a space-age museum.

● Left: *Astounding detail is visible in this view of Europe and North Africa without clouds. It is a composite of images taken by NOAA weather satellites, processed to give near-natural colours (except for urban areas, which are shown in red).*

The planet studied most thoroughly from space is our home, the Earth. Every day, satellites photograph the Earth's surface and its weather. Most familiar are the pictures that appear on TV weather bulletins. These come from two types of satellite, distinguished by their orbits. Satellites in *geostationary orbit*, 36,000 km (22,300 miles) above the equator, photograph the Earth every half-hour. (At this distance a satellite orbits the Earth in exactly one day, and so remains above a particular spot on the equator.) The other type of satellites orbit from pole to pole at lower altitudes, and so can photograph an area in greater detail but at longer intervals.

As well as simple views of cloud coverage and weather fronts, the sensors on these satellites reveal the temperature of land and sea, the height of clouds, and the humidity of the atmosphere. Such information improves forecasts, allows meteorologists to make long-range predictions, and even saves lives and property by giving advance warning of major storms, which is of particular importance in areas prone to hurricanes and flooding.

Earth resources satellites, such as the US Landsats, are more concerned with the surface of the planet, although their role somewhat overlaps that of the weather satellites. Orbiting from pole to pole, they cover the entire globe as the Earth spins beneath them. They take photographs in several different regions of the spectrum, including the infrared. The pictures are computer-processed and often given false colours to emphasize certain details such as vegetation or types of rock.

Satellite pictures show the Earth's immense beauty and variety, but also reveal how human habitation is changing and threatening the planet. Scientists use Landsat pictures to study many aspects of our environment including agriculture, forestry, the oceans, water and air pollution, acid rain, and the spread of urban development. Geologists can search for new sources of minerals, map faults that produce earthquakes, and see the results of volcanic eruptions. Weather and Earth resources satellites keep watch on our planet far more effectively than could be done in any other way, and they will play a vital role in monitoring such effects as climatic change and the hole in the ozone layer.

In addition, spy satellites regularly photograph the Earth in search of military and economic information on other countries. Their immensely powerful cameras can see details as small as 15 cm (6 inches) across. Such satellites are essential for verifying arms control agreements.

● *As well as observing in visible light, Meteosat weather satellites observe in the thermal infrared (below right), measuring temperatures of clouds and of the surface of the Earth, and also at the wavelength which is absorbed by water vapour in the atmosphere (below left).*

129

● *North and South America from the*
GOES-E geostationary satellite.

● *Africa and Europe from the Meteosat 2 geostationary satellite.*

● *Earthrise, seen from orbit around the Moon by the Apollo 11 astronauts.*

● Left: *A mosaic of cloud-free pictures of Antarctica taken by NOAA weather satellites.*

Looking back from orbit around the Moon, the crew of Apollo 8, a pathfinder flight for the first manned landing, became the first people in history to see the Earth as, in their words, "a grand oasis in the vastness of space." What is it that makes the Earth, of all the planets in the Solar System, so eminently suitable for life, and what can this tell us about the possibility of life elsewhere in the Universe?

The first requirement for life in any planetary system is a stable, long-lived star. Our Sun is just such a star, having shone steadily for 4.6 billion years and with billions more years in prospect. If the Sun's brightness had been highly variable then conditions on Earth would have been too unstable for life to have arisen. The second requirement for life is a suitable planet at the right distance from the parent star. The Earth, third planet from the Sun, fits that bill admirably.

Around every star is an imaginary 'green belt' in which a planet will have the right surface temperature for water to exist in liquid form. Water is essential to life as we know it, and the abundance of liquid water on Earth makes it uniquely favourable for life in our Solar System. Had the Earth been closer to the Sun it would have overheated and all the water would have evaporated, as has happened on Venus. Had it been farther from the Sun it would now be locked in a permanent ice age, like Mars.

The Earth's current average surface temperature is around 14°C, although this has varied in the past and will probably do so again. For comparison, the average surface temperature of Mars is −55°C. However, Mars would have been warmer if it had been a larger planet since its increased gravity would have allowed it to retain a denser atmosphere, so trapping more heat. Hence a simple matter like size can also affect a planet's suitability for life.

Seen from space, the Earth appears as a blue and white ball with an equatorial diameter of 12,756 km (7926 miles). The blue parts are the oceans. In fact, nearly three-quarters of our planet is covered with water, so the name 'Earth' is rather inappropriate. Additional water is locked up in frozen form in the sparkling polar ice caps and as snow on mountain ranges. Much of the surface is obscured by white clouds of water vapour and ice crystals, at an altitude of up to 15 km (9 miles) in the case of the highest cirrus.

Those swirling clouds are unmistakable evidence of a plentiful atmosphere, essential for life. Nitrogen (77 per cent) and oxygen (21 per cent) are the main constituents. No other planet in the Solar System has such a high proportion of atmospheric oxygen, because no other planet possesses life. The oxygen in our atmosphere was released by the breakdown of carbon dioxide by plants. Without it, oxygen-breathing organisms such as ourselves could never have evolved.

Stars brighter and hotter than the Sun have larger green belts, which increases the chances of finding a suitable planet. But brighter stars also burn out quicker, which gives intelligent life less of a chance to develop.

● The 'green belt' around the Sun within which temperatures are suitable for life extends from the orbit of Venus to beyond the orbit of Mars. Other stars have different-sized habitable zones, depending on their size and temperature.

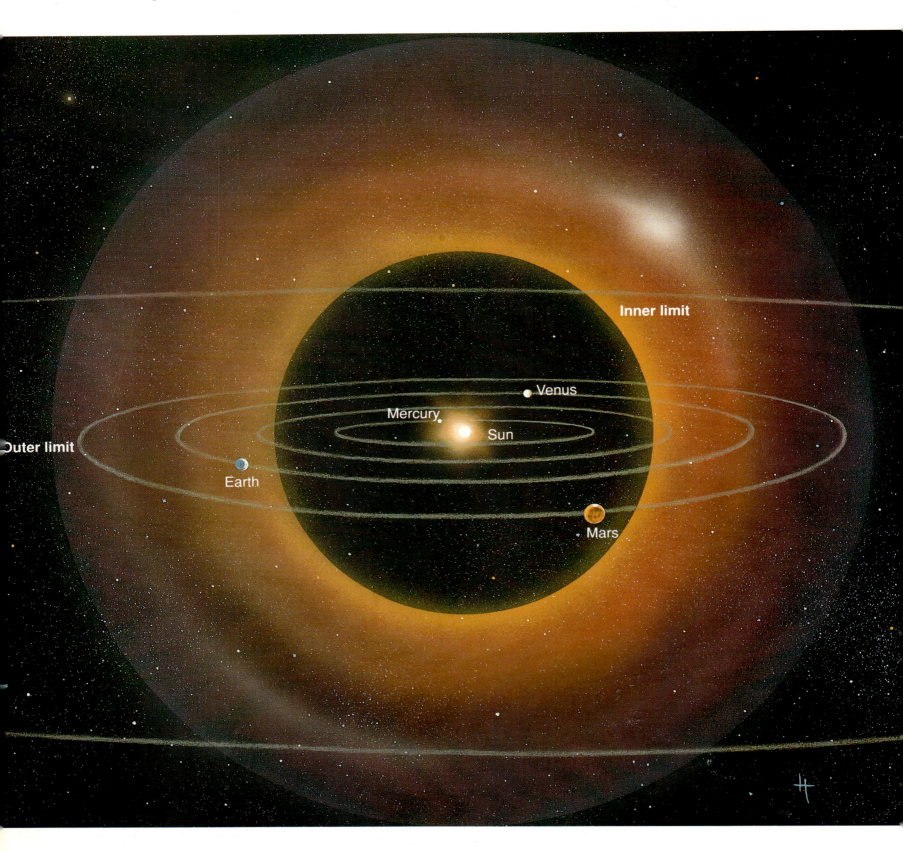

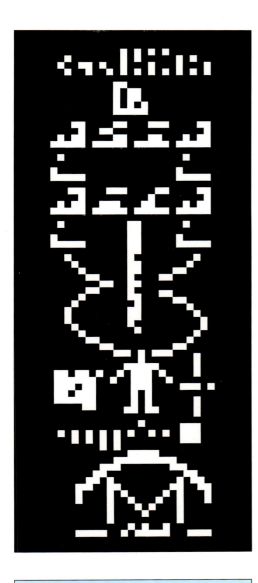

THE ARECIBO MESSAGE

On November 16, 1974 a signal was sent from the world's largest radio telescope, 305 metres (1000 feet) in diameter, at Arecibo, Puerto Rico, towards the 300,000 stars of the globular cluster M13 in Hercules. The message consisted of 1679 on/off pulses that can be arranged into the picture shown above, giving information about the chemical ingredients of terrestrial life, the shape and size of a human, the planets of the Solar System, and the Arecibo radio dish itself. The Arecibo message lasted three minutes and was sent only once. Rather than a serious attempt at interstellar signalling, the intention was to put together an example of what we may one day hear if there is anyone out there trying to signal to us.

● *Artist's impression of hypothetical life-forms in the atmosphere of a giant, gaseous planet such as Jupiter.*

The stars with the longest lifetimes are the faint red dwarfs, whose green belts are so narrow and close in that the chances of a suitable planet orbiting them are slim. On all counts, stars like our Sun are ideal ones around which to find life.

But do planetary systems exist around other stars? Planets of other stars would be too faint to observe directly, but astronomers have looked for indirect signs of their existence, particularly for slight irregularities in the motion of their parent stars through space. A star with accompanying planets would not move in a straight line, as would a lone star, but would weave slightly from side to side as the planets orbited it. The movements are small and difficult to detect, but astronomers are starting to amass evidence that may confirm the existence of other planetary systems.

Perhaps one star out of every ten has planets – and since there are several hundred billion stars in our Galaxy, that implies the existence of a lot of planets out there. Of course, not all these stars are of the stable, long-lived type like the Sun, and not all will have a suitable planet in their green belt. Even so, there could be millions of planets throughout the Galaxy where life might exist.

What would alien life look like? In many places, life might not progress beyond the stage of simple plants. Other planets may be dominated by ferocious creatures like the dinosaurs that once ruled the Earth. Life will evolve to suit a particular planet. On larger worlds with strong gravity and a dense atmosphere, life-forms would be squat and sluggish, whereas a smaller, lower-gravity world would favour slender, more elegant creatures. Giant, gaseous planets like Jupiter, for example, might harbour life-forms that float in their clouds.

Life first arose on Earth over 3000 million years ago but it is only in the past century, with the development of radio, that we have become able to signal our existence to the cosmos. Modern radio telescopes are so powerful that they are capable of communicating with similar equipment anywhere in the Galaxy. In the hope that there is already someone out there trying to attract our attention, astronomers have attempted to detect incoming radio messages, but so far without success. However, the Galaxy is so big, and there are so many possible radio frequencies to choose from, that much more searching needs to be done before we can tell whether there is intelligent life elsewhere in the Universe.

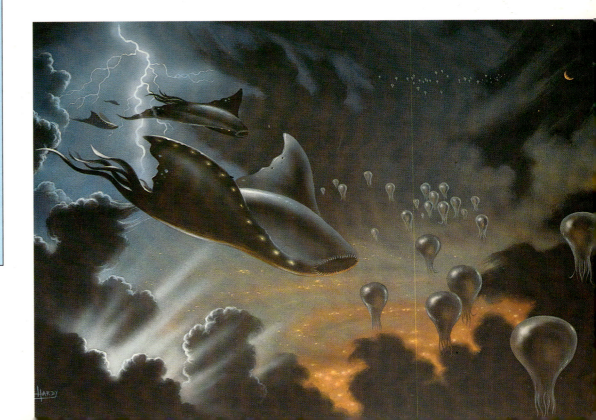

The Earth is continually peppered by natural debris from space. Most of it arrives as shooting stars – a complete misnomer since they are nothing whatever to do with stars. They are actually dust particles from comets and asteroids, burning up by friction high in the atmosphere, and they are more correctly termed *meteors*. A typical meteor weighs no more than a gram. What we see is not the particle itself, but the streak of incandescent gas it produces as it plunges into the atmosphere at high speed about 100 km (60 miles) above our heads.

On any clear night, away from city lights, five or six meteors should be visible to the naked eye each hour, entering the atmosphere at random. These are known as *sporadic* meteors. At certain times of the year the Earth enters dust lanes spread along the orbits of comets (and, in one case, an asteroid). The result is a meteor shower, in which dozens of meteors can be seen each hour apparently streaming away from a single point in the sky known as the *radiant*. A shower is named after the constellation in which its radiant lies (or sometimes a bright star near the radiant): for example, the brightest shower of the year, the Perseids, appears to radiate from Perseus, while another fine shower, the Geminids, comes from Gemini.

At best, several dozen meteors per hour can be seen during busy showers such as the Perseids or Geminids, an average of roughly one a minute. Much rarer are spectacular *meteor storms*, when thousands of meteors per hour can be seen cascading from the heavens. One such storm was seen over North America on the night of November 17, 1966. A similar storm occurred in 1833, when observers described meteors as falling like snowflakes. Both storms were caused by the Leonid meteors, shed from Comet Tempel–Tuttle. The distribution of meteors along the comet's path is patchy and leads to occasional bursts of activity at 33 year intervals, the comet's orbital period. Astronomers will be watching for another major Leonid meteor storm on November 17, 1999.

● Below left: *Several times a year the Earth passes through swarms of dust orbiting the Sun, producing meteor showers. Pictured here is a bright meteor (the long, almost vertical streak) from the Geminid shower that occurs each December. The other, shorter trails are stars.*

There is no chance of an ordinary meteor reaching the ground since the particles are too small and fragile. However, from time to time the Earth encounters much more substantial chunks of rock or metal which fall to Earth as *meteorites*. They are thought to have a different origin from meteors – most of them fragments of asteroids rather than comets. Some meteorites are slowed so much by their passage through the atmosphere that they drop harmlessly to the ground, but the largest ones are still moving so quickly that they vaporize on impact, blasting out a huge crater many times their own diameter.

In the Arizona desert near the town of Winslow lies a crater 1.2 km (0.75 miles) in diameter, misleadingly called Meteor Crater – the correct name should be Meteorite Crater, for it was caused by the impact of an iron meteorite that is estimated to have weighed around a quarter of a million tonnes. It is a smaller version of the craters that scar the surface of the Moon, Mercury, and many other bodies in the Solar System. Most of the Arizona meteorite was destroyed in the impact, and only a few fragments remain. The largest meteorite to have survived lies where it landed, at Grootfontein in Namibia; it, too, is composed of iron and is estimated to weigh about 60 tonnes.

● *'Meteor Crater' in the Arizona desert, formed in prehistoric times by the fall of an immense iron meteorite.*

Although probably thousands of meteorites hit the Earth each year, most of them go unnoticed since they are small and fall in the ocean or in uninhabited regions. Occasionally though, a meteorite causes some damage. On November 30, 1954, Mrs Ann Hodges of Sylacauga, Alabama, was struck on the hip by a small meteorite that crashed through the roof of her house. She is the only person in recent times known to have been injured by a meteorite, but a number of buildings have been damaged.

Meteorites are of interest since they provide us with samples of extraterrestrial material. As mentioned above, most meteorites are thought to be fragments of asteroids, although some fragile stony meteorites are thought to have come from the heads of comets. In recent years scientists have found large numbers of meteorites preserved in deep-freeze in the ice of Antarctica where they fell long ago. By contrast, meteorites that fall on other parts of the Earth are soon weathered away. Among the Antarctic meteorites are some unusual rocks whose origin is of special interest – they are believed to be pieces from the Moon and Mars, blasted off by ancient impacts.

● *A meteorite collected in Antarctica, where it had been preserved in natural deep-freeze. Most meteorites are thought to be fragments from asteroids. However, this meteorite is one of a rare group thought to come from the Moon. It is shown here actual size.*

● Above left: *Trees were felled like matchsticks for many kilometres around the site in Siberia where a comet plunged to Earth in 1908. This photograph was taken 22 years after the event.*

● Above right: *A 1963 photograph of the Tunguska site showing the new growth of vegetation through the felled trees.*

Early on the morning of June 30, 1908, a huge fireball blazed through the sky and exploded with the force of a nuclear bomb over the valley of the Tunguska river in Siberia. The blast flattened trees for up to 30 km (20 miles), and a farmer sitting on his porch 65 km (40 miles) away was knocked unconscious. Heat from the explosion melted metal objects, incinerated reindeer, and set the forest ablaze. Several hundred kilometres to the south, the driver of the Trans-Siberian express stopped his train as the rails and carriages began to shake. Seismic waves like those from an earthquake were recorded throughout Europe. It was a natural disaster rivalled only by the eruption of the volcanic island of Krakatoa in 1883.

Scientists now believe that the Tunguska blast was caused by the collision with the Earth of a small comet about 100 metres (300 feet) in diameter, possibly a fragment from the head of Encke's Comet. Dust from the object produced eerie bright nights throughout Europe for days afterwards.

DID AN ASTEROID KILL THE DINOSAURS?

What happened, 65 million years ago, to wipe out the dinosaurs? At that time there was an environmental disaster of global proportions which caused a mass extinction — not just of the dinosaurs, but of half the species of living things including many other land and sea creatures.

Recently suspicion has focused on an impact by a large asteroid, which could have caused a major change in the Earth's climate. Evidence for the impact comes from a layer of clay laid down around the world 65 million years ago, presumed to have formed from dust from the impacting body. In its composition the clay is not at all like Earth rocks, but is similar to that of rocky meteorites believed to be fragments of asteroids.

If an asteroid, or possibly a comet, 10 km (6 miles) across, had collided with the Earth, copious amounts of dust and perhaps water vapour would have been flung up and spread throughout the atmosphere, drastically reducing the amount of sunlight reaching the surface. This would have radically affected the climate for many months or even years, and many species would have died out by the time the dust settled to form the layer of clay.

It has been suggested that the location of the impact was in the Caribbean, south of Cuba, where the 65-million-year-old deposits are thickest. Ominously, a 10 km diameter asteroid is expected to hit the Earth every 100 million years or so. Next time it will be humans that face extinction.

6 GALAXIES AND THE UNIVERSE

It is summer again, and on a Moonless night the Milky Way cleaves the sky like the luminous wake of a ship. "Gala," says my young companion, using the Greek name for it, meaning 'milk', from which come our terms Milky Way and Galaxy. "The infant Heracles, the illegitimate son of Zeus by a mortal woman, was placed at the breast of the goddess Hera while she slept, so that he might suckle her milk and become immortal," she relates. "But Hera awoke and pushed him away, and her milk squirted across the sky."

That, at least, was the myth. Democritus, the Greek best known for popularizing a theory of atoms, is credited with originating the idea that the Milky Way is a mass of faint stars, but it was not until 1609, when Galileo turned his telescope on the Milky Way, that Democritus was proved correct. A pair of binoculars will reveal billowing star clouds along the Milky Way, becoming particularly dense in the regions of Sagittarius and Scorpius, the direction of our Galaxy's centre.

In fact, the Milky Way *is* our Galaxy, and the two terms are used interchangeably. All the stars that comprise the constellations are members of our Galaxy, only much closer to us than the stars that mass along the band of the Milky Way. The very existence of the Milky Way is a clue to the shape of our Galaxy, since it shows that the distribution of stars is not random. In 1784 Sir William Herschel counted the stars visible in various directions through his telescope and concluded that the Galaxy was shaped rather like a grindstone, about five times as long as it is broad, with the Sun situated somewhere near the centre.

This view prevailed until 1918 when Harlow Shapley, an American astronomer who was working at Mount Wilson Observatory outside Los Angeles, California, made an advance that in its own way was as significant as when Copernicus dethroned the Earth from the centre of the Solar System (see p. 62). Shapley discovered that not only was the Galaxy much bigger than anyone had suspected, but the Sun was nowhere near central.

Shapley's discovery emerged from his studies of globular clusters, giant ball-shaped aggregations of stars (see p. 143). Shapley noted that globulars are concentrated in one half of the sky, centred on Sagittarius. He made the bold assumption that the globulars were actually arranged symmetrically in a halo around our Galaxy, and that the apparent asymmetry is a consequence of us looking from a vantage point some way towards the Galaxy's edge.

● *Tintoretto's* Juno and the Infant Hercules *depicts the mythical origin of the Milky Way (Juno was the Roman name for the Greek goddess Hera).*

● Left: *Stars mass together in the Milky Way in Sagittarius, towards the centre of our Galaxy. At the top is M17, the Omega Nebula.*

● Right: *The globular cluster M55 is a ball-shaped swarm of stars in Sagittarius. Its brightest members are red giants.*

GLOBULAR CLUSTERS

Scattered in a corona around our Galaxy are 140 or so ball-shaped flocks of stars, all slowly orbiting the Galaxy's centre, in much the same way as long-period comets orbit the Sun. A typical globular cluster is 100 light years in diameter, and the largest contain up to a million stars — they are virtually mini-galaxies. Yet to the naked eye the clusters are barely visible, they are so distant.

Omega Centauri and 47 Tucanae are the largest and brightest globulars as seen from Earth, both looking like faint, swollen stars, but they are too far south to be visible from mid-northern latitudes. The best globular for northern observers is M13 in Hercules, just visible to the naked eye on clear nights as a chalky smudge. It lies 25,000 light years from us.

Globular clusters contain some of the oldest stars known, at over 10,000 million years, and they must have been among the first parts of our Galaxy to form, presumably while it was still flattening into its disk shape. All galaxies worth their salt have a complement of globular clusters. If we lived in a globular cluster the sky would be studded with stars brighter than the planet Venus, and the 'night' would never be dark.

Shapley worked out distances to the globulars from variable stars in the nearer ones and the brightest stars in the more distant ones. His results suggested that the Galaxy was ten times larger than previously supposed, and that the Sun lay 50,000 light years off-centre. Previous astronomers had been misled into thinking the Galaxy was smaller because dust in space obscured our view of the most distant stars.

Modern work has revised Shapley's figures downwards somewhat, although his basic conclusion still holds. The Galaxy is now estimated to be 100,000 light years in diameter, with the Sun a suburban star about two-thirds of the way to the rim. The galactic centre lies around 30,000 light years away in Sagittarius, near that constellation's borders with Ophiuchus and Scorpius. According to current estimates, our Galaxy contains upwards of 100,000 million stars.

But what, if anything, lies beyond the Galaxy? At first, Shapley believed that the Galaxy was surrounded by nothing but empty space – that in fact, for all practical purposes, the Galaxy *was* the Universe. However, this did not solve the problem of the so-called *spiral nebulae*, misty patches that many other astronomers believed to be extremely distant whorls of stars.

The idea that nebulae might be separate Milky Ways was not new. William Herschel had charted many nebulae during his surveys of the sky in the late eighteenth century, and referred to them as 'universes'. Lord Rosse, an Irish astronomer, studied the nebulae through a giant telescope of his own construction which had a mirror 1.8 metres (72 inches) in diameter, making it the largest instrument of its day. With the increased light grasp of this telescope, Rosse was able to detect spiral structure in some nebulae, notably in the object listed by Messier as M51, and now also known as the Whirlpool. He drew it in 1845 looking like a nautilus shell.

● Above: *Harlow Shapley, who showed that the Sun lies in the suburbs of the Galaxy.*

● Below: *Lord Rosse, the Irish astronomer who in 1845 discovered the spiral nature of galaxies with his giant telescope (below left), which was suspended between high walls in the grounds of his castle.*

Analysis of the light from various nebulae by spectroscopy (see p. 43) settled the question of their nature, although not their distances. In the 1860s the English scientist William Huggins discovered that there were two types of nebula, distinguishable by their spectra. While some nebulae, such as the one in Orion, have spectra that identify them as gaseous, the spectra of others, including the spiral nebulae, confirm that they are composed of stars.

However, individual stars could not be discerned in spiral nebulae, even in telescopes as powerful as Rosse's. And without the brightnesses of any individual stars to measure, distances could not be calculated. Shapley believed the spirals to be either within an enlarged Galaxy or, at the very most, just beyond it. Already, though, there was evidence to prove him wrong in the form of novae that had been photographed in a few spiral nebulae, and the final proof was not long in coming.

Even as Shapley was completing his work on the size of the Galaxy, a new telescope larger than any in the world was beginning operation on Mount Wilson. It had a mirror 2.5 metres (100 inches) across and is still colloquially known as the 100-inch. In the hands of Edwin Hubble, this telescope was to make a series of momentous discoveries that underpin our present-day understanding of the Universe.

Hubble began by photographing M31, the great spiral nebula in Andromeda, to find Cepheid variable stars in it. To the naked eye, M31 is visible in a clear sky as an elongated smudge spanning more than the width of the full Moon. Once Hubble had discovered Cepheids within it he could estimate its distance – and it turned out to be way beyond the limits of even Shapley's enlarged Milky Way. The spiral nebulae were other Milky Ways after all, separated by immense gulfs of space. Shapley capitulated, and popularized the term 'galaxies' for them. (When we speak of our own Galaxy we use a capital letter to distinguish it from any other galaxy.) Hubble's discovery threw open the doors on an unimaginably vast Universe.

● Above: *Edwin Hubble, the discoverer of the expanding Universe.*

● Below: *Lord Rosse's drawing which showed the spiral shape of M51, now known as the Whirlpool Galaxy, as seen through his giant telescope. Even the satellite galaxy at the end of a spiral arm is recorded. (Compare this drawing with the photograph shown on p. 155).*

To see how our Galaxy would appear from outside, we have only to look at other spirals such as the Andromeda Galaxy. A spiral galaxy has the overall form of a discus. Side-on it appears elongated with a raised central hub, like NGC 4565. Our Galaxy's hub is about 12,000 light years thick but the thickness tapers towards its rim, reaching about 2000 light years in the Sun's vicinity. A stratum of dust and gas lies in the plane of the Galaxy; such lanes are particularly noticeable in NGC 4565 and also in the Sombrero Galaxy.

In galaxies seen face-on, like NGC 2997, pink nebulae and blue clusters of bright young stars are strung like coloured beads along the spiral arms, for this is where star formation takes place. Two (or possibly more) arms of stars, gas, and dust wind outwards like coils of rope from our Galaxy's central boss. Branches sprout from each spiral arm so that, at the distance of the Sun, the structure of the Galaxy is quite complicated.

● Right: *A beautifully coiffured spiral galaxy, NGC 2997 in Antlia. Pink nebulae, sites of star formation, are dotted among the young, blue stars of its spiral arms.*

● Left: *NGC 4565 in Coma Berenices is a spiral galaxy seen edge-on with a dark lane of dust running through it. Viewed from outside, our Galaxy would look like this.*

● Right: *The Sombrero Galaxy, M104 in Virgo, is a spiral seen almost edge-on with a large central bulge containing old, red stars. Younger stars produce the blue colour in the outer regions. Its spiral disk froths with gas and dust clouds, seen in silhouette across the nucleus. Numerous globular clusters can be seen as diffuse spots around the Sombrero; hard, rounded images are foreground stars in our Galaxy.*

From our position in the thick of things it is difficult to identify the arms of our Galaxy, but radio astronomers have traced their general pattern by mapping the hydrogen gas within them. Hydrogen, the most abundant substance in the Universe, emits radio waves at a wavelength of 21 cm which cut through the dark clouds of dust that obscure our view in visible light. The rotation of the Galaxy causes slight changes in the wavelength of the 21 cm emission by the Doppler effect (see p. 157), allowing radio astronomers to trace the motions and positions of the emitting hydrogen clouds.

All galaxies are rotating at a rate that is imperceptible in a human lifetime, although over many centuries the rotation of our Galaxy slowly distorts the familiar constellation shapes. As in any orbiting system, objects nearest the centre move the quickest. Our Sun takes about 200 million years to complete its orbit around the galactic centre, which means that the Earth has made about 20 such orbits since it was born.

The neighbouring spiral arms of our Galaxy can be picked out by the prominent nebulae and open clusters they contain. Towards the galactic centre lies the so-called Sagittarius Arm, which contains such familiar sights as the Lagoon Nebula and the Trifid Nebula in the constellation Sagittarius and the Eagle Nebula in nearby Serpens. This arm continues into Carina where we are looking along the barrel of it, and includes Eta Carinae. Facing away from the galactic centre, we see the Perseus Arm,

● Below: *The Large Magellanic Cloud, a satellite galaxy of our Galaxy, lies about 160,000 light years away and is visible to the naked eye in southern skies. Because of its proximity, the cloud is easily resolved into star clusters and nebulae. The large pink area at centre left is the Tarantula Nebula, the largest nebula known, near which Supernova 1987A appeared.*

● Right: *The Andromeda Galaxy, M31, a larger version of our Milky Way, lies over 2 million light years away and is the most distant object visible to the naked eye. Its central hub is like a pale egg yolk, tinted by the light of old, cool stars. Two satellite galaxies are visible, one (M32) just left of centre and the other (M110) below right.*

containing the Double Cluster and the Crab Nebula. Both the Sagittarius and Perseus Arms are about 6000 light years away. Astronomers think that we lie on an inward branch of the Orion Arm, the main part of which contains the Orion and Rosette Nebulae. (Photographs of these nebulae appear in Chapter One.)

Our Galaxy is orbited by two small satellite galaxies, called the Magellanic Clouds after the Portuguese explorer Ferdinand Magellan, whose expedition brought news of them to Europe. To the naked eye they look like splashes from the Milky Way, close to the southern celestial pole. The Large Magellanic Cloud, the nearer of the two at 160,000 light years, spans the width of an outstretched hand in the constellation Dorado. It is elongated in shape, about one-third the diameter of our Galaxy at its longest, and contains about a tenth as many stars. Its most prominent feature is the spidery-shaped Tarantula Nebula, near which a bright supernova erupted in 1987 (see Chapter One).

About 30,000 light years farther off lies the teardrop-shaped Small Magellanic Cloud, half the width of its larger relative and stocked with only a fifth the number of stars. Recent work suggests that the Small Magellanic Cloud is in fact an elongated stream of stars presented end-on, the result of a near collision 200 million years ago with the Large Cloud. The encounter disrupted the original Small Cloud, which is now dispersing into space and will eventually disappear.

The most distant object that the unaided eye can see is the Andromeda Galaxy, M31, which modern research places at a distance of 2.4 million light years. In other words, the light now reaching the Earth left there over 2 million years ago, when ape-men roamed the plains of Africa. The Andromeda Galaxy is the Milky Way's big brother, about half as wide again and containing perhaps twice as many stars. Like our Galaxy it has a spiral structure, not immediately obvious since it is tilted at a steep angle to us, but there are at least two spiral arms, traced out by dark lanes of dust silhouetted against the nucleus. The Andromeda Galaxy even has its equivalents of the Magellanic Clouds – M32 and M110, two satellite galaxies that can be seen in small telescopes.

● *The Local Group of galaxies, a loose cluster of which our Milky Way is a major member. The galaxies in the group are held together by gravity as they wheel through space. The lines in the reference grid are approximately 650,000 light years apart, and the vertical lines show distances above and below the grid.*

Key to numbers
(clockwise from Milky Way at centre):

1 Milky Way Galaxy
2 Sculptor
3 Small Magellanic Cloud
4 Large Magellanic Cloud
5 Fornax
6 Carina
7 IC 5152
8 Sextans A
9 Leo A
10 Leo I
11 Sextans C
12 Leo II
13 Ursa Major
14 Draco
15 Ursa Minor

16 Pegasus
17 NGC 6822
18 Wolf-Lundmark
19 IC 1613
20 LGS3
21 DDO 210
22 Andromeda I
23 Andromeda III
24 M32 (NGC 221)
25 M110 (NGC 205)
26 Triangulum, M33 (NGC 598)
27 Andromeda, M31 (NGC 224)
28 NGC 185
29 Maffei I
30 NGC 147

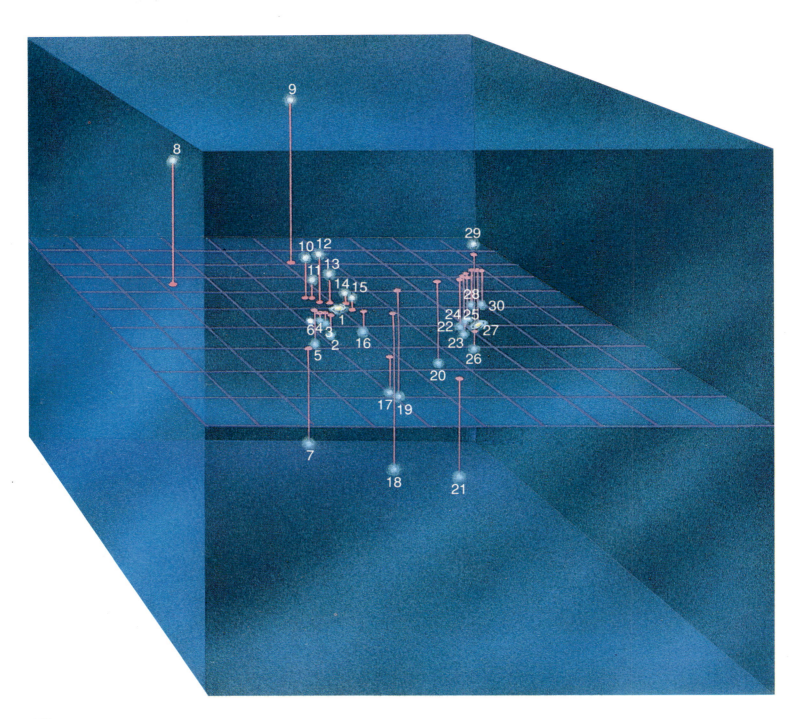

150

The Andromeda Galaxy and our Milky Way are the two largest members of a cluster of galaxies known as the Local Group. About 30 members of the Local Group are known, bound together by gravity in something like a larger version of an open star cluster. The only other member of note is M33, a spiral galaxy in Triangulum, rather more distant than M31 but not much larger than the Large Magellanic Cloud. All the other members of the Local Group are smaller still, and most of them are concentrated around either the Milky Way or M31. There are probably many more small, faint members of the Local Group still to be discovered.

Spirals like M31 and the Milky Way are not the only species of galaxy, as Edwin Hubble found when he probed into the Universe with the Mount Wilson telescope. He classified galaxies according to their shapes, and arranged the various types into the so-called *tuning-fork diagram* (see pp. 152–153). This diagram gives a false impression of continuity, with the implication that galaxies may evolve from one type into another, but in fact they do not.

In some·spirals the arms stem from a straight bar of stars and dust that crosses the central hub. Hubble termed these *barred spirals* and classified them as type SB; ordinary spirals are type S. Both types are subdivided further, by appending lower-case letter a, b, or c according to how tightly their arms are wound. The Andromeda Galaxy is an Sb spiral. Our own Galaxy, whose arms are less tightly wound, is thought to be midway in type between Sb and Sc. However, some astronomers think that there may be a short bar across the centre of our Galaxy, which would mean that it is actually a type SB.

Along the handle of the tuning fork Hubble placed the *elliptical galaxies*, type E, which range in shape from the perfectly spherical, E0, to the most elliptical, E7. Unlike spirals, elliptical galaxies consist mostly of old stars and contain virtually no dust or gas for producing new ones. Galaxies of type S0, intermediate between ellipticals and spirals, are even more flattened than E7 galaxies; they may be former spirals that have somehow been stripped of their arms.

The irregulars, designated Irr, do not fit comfortably into any of the other categories. Both Magellanic Clouds are classified Irr, although the Large Cloud's structure is somewhat suggestive of a barred spiral. Only a few per cent of galaxies are irregulars.

Elliptical galaxies show the widest range in size of all types. Supergiant ellipticals are the largest galaxies known, containing as many stars as 100 Milky Ways, but they are rare. At the other end of the scale, dwarf ellipticals may contain only a few million stars; they are little more than large globular clusters. Dwarf ellipticals are probably the most numerous galaxies in the Universe, but they are difficult to see because they are so faint. Most of the galaxies in the Local Group are dwarf ellipticals.

Spiral galaxies have a more limited size range. The Andromeda Galaxy is about as large as spirals get, while M33 is a typical small spiral. Ordinary spirals outnumber barred spirals by about two to one. There are no barred spirals in the Local Group (unless our own Galaxy turns out to be one).

Why some galaxies should be elliptical and others spiral continues to be a puzzle for astronomers. However, it is thought that galaxies formed from giant collapsing clouds of hydrogen gas, and that the clouds spinning the fastest as they shrank became spirals, whereas the slowest-spinning clouds became ellipticals.

Most galaxies belong to clusters ranging from small, haphazard groupings of only a handful of members, smaller even than our Local Group, to congregations of thousands which are often centred on a supergiant elliptical galaxy. Scattered in the northern part of the constellation Virgo and over the border into Coma Berenices are

● Centre: *Hubble's 'tuning fork' diagram, a classification of galaxy types. To the left are ellipticals, type E. Spirals are split into ordinary types, S, and barred spirals, SB, and are subdivided according to how tightly their arms are wound. Despite the impression this diagram gives, galaxies do not evolve from one type into another.*

● Right: *M83 in Hydra is a spiral with two prominent swept-back arms, filled with dark dust lanes and bright nebulae.*

● Far right: *M101, a large Sc type spiral in Ursa Major with outflung arms.*

● Bottom right: *A classic barred spiral galaxy, NGC 1365 in Fornax. The colours show the clear division into older, cooler stars in the galaxy's centre and young, blue stars in the spiral arms.*

● Bottom centre: *The Small Magellanic Cloud, an irregular galaxy that is a satellite of the Milky Way, with the globular cluster 47 Tucanae to its right.*

● Below: *M87, a giant elliptical galaxy in the Virgo cluster, surrounded by hundreds of globular clusters. M87 is a strong radio source, and shorter time exposures show a jet of gas apparently being shot out from the galaxy's core.*

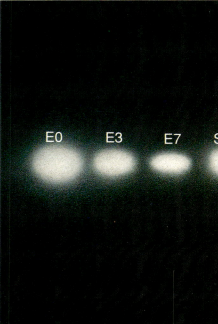

E0   E3   E7   S

Sb

Sc

Sa

Irregular

SBa

SBb

SBc

thousands of galaxies, members of the Virgo Cluster, the nearest large cluster to us at about 50 million light years away. Its brightest members are the ellipticals M49 and M87, the latter also known as the radio source Virgo A. Strong radio emission from a galaxy is a sign of turmoil, and photographs show a jet of gas squirting from the core of M87 at high speed as though being slung out by some violent event.

In the background in Coma Berenices, just north of the Virgo Cluster but over 200 million light years more distant, is another cluster of galaxies quite different in character. The Coma Cluster contains mostly elliptical and S0 galaxies, whereas large spirals predominate in the Virgo Cluster. Just as individual galaxies are found in clusters, so do clusters of galaxies seem to group together, in *superclusters*, in a dizzying hierarchy throughout the Universe. The Local Supercluster, of which the Local Group is a fringe member, is centred on the Virgo Cluster.

● Below: *The central portion of the Virgo cluster of galaxies, showing several spirals and ellipticals. The brightest members here are M86 (centre), an S0 galaxy, and M84 (right), a giant elliptical.*

● Top: *This ring-shaped galaxy about 300 million light years away in the constellation Volans is about 150,000 light years in diameter, as large as the Milky Way. Its shape is thought to have been caused by another galaxy plunging through it.*

● Above right: *A companion galaxy which seems to lie at the end of a spiral arm of M51, the Whirlpool Galaxy, actually lies behind the arm, having once brushed past it.*

● Above: *The Antennae, a pair of galaxies tidally distorted by their close approach to produce long 'feelers' of stars and gas millions of light years long.*

Occasionally galaxies brush past each other, collide, or even merge. These encounters are the most spectacular traffic accidents in the Universe. An example is the famous Whirlpool Galaxy, M51. Astronomers had known since the days of Lord Rosse that a smaller companion sits at the end of one of its spiral arms, but only recently have they come to realize that the smaller companion is moving past M51 almost directly away from us – it actually lies behind the spiral arm that it appears to be peeling away from the larger galaxy. The two galaxies are no longer in contact, although they undoubtedly touched a few hundred million years ago.

Another elegant galaxy–galaxy interaction is known as the Antennae because of its resemblance to the feelers of an insect. Here, two spiral galaxies of comparable size have waltzed past each other over hundreds of millions of years, their gravities pulling out a tail of stars and gas from each other. Again, the two are not touching, although they seem to be from our perspective. Perhaps the oddest-looking galaxies of all are the ring galaxies, believed to be spirals which a small galaxy has passed clean through, setting off a ripple of star formation.

Centaurus A, a strong radio source known also as NGC 5128, is apparently the result of a merger between two galaxies. At first sight it appears to be an elliptical galaxy girdled by a dust lane. But ellipticals do not have dust and gas – that is a characteristic of spirals (as we have seen). It seems, therefore, that a large elliptical galaxy has welcomed a smaller spiral into its embrace, and devoured it, sparking off a round of star formation. These young stars show up blue on colour photographs.

If two spirals meet and merge they will produce an elliptical, and some elliptical galaxies are now believed to have formed in this way. There is evidence that the supergiant ellipticals found at the centres of large clusters of galaxies have grown to their present size by the merger of several smaller galaxies, a process some astronomers have termed galactic cannibalism.

● *The radio source Centaurus A is a strange-looking supergiant galaxy that may have resulted from a merger between an elliptical and a spiral.*

As Hubble explored ever deeper into space with the 100-inch on Mount Wilson, he found more and more galaxies stretching as far as the telescope could see. He was observing not by looking through the telescope, but by making long time exposures on photographic plates. The longer the exposures, the more galaxies showed up.

Hubble measured the distances to these galaxies first by Cepheid variables for the closest ones, then by the brightest stars in those too distant to show individual Cepheids and finally, when even the brightest stars were out of range, by the overall brightness of the galaxy. At the same time, he and other astronomers were measuring the motion of each galaxy from the displacement of lines in their spectra caused by the Doppler effect. They found that galaxies are not fixed in space but move at high speeds relative to the Milky Way.

The motions are not random, though. Apart from the closest galaxies such as the Andromeda spiral, all galaxies are moving away from us, as is revealed in the *red shift* of their spectra (see the box). Hubble's astounding discovery, announced in 1929, was that the red shift – and hence the speed of recession — increased with distance. In other words the Universe was expanding, like a balloon being inflated. Coming only 10 years after Shapley's reassessment of the size of our Galaxy, the discovery of the expansion of the Universe was the culmination of the most astounding decade in modern astronomy.

The relationship between red shift and distance is known as *Hubble's law*, and the factor linking a galaxy's distance to its speed of recession is *Hubble's constant*. Red shifts are easy enough to measure but distances to galaxies are not, and so the exact value of Hubble's constant remains uncertain. Estimates range between 50 and 100 kilometres per second per million parsecs (the units in which astronomers traditionally express it); most astronomers settle for a compromise figure of 75.

It is important to realize that it is only the space between clusters of galaxies that is expanding, not the clusters themselves or the galaxies they contain. In clusters such as the Local Group, the mutual gravitational attraction between galaxies overrides the expansion. The Andromeda Galaxy is actually approaching us because it and our Galaxy are gravitationally tied, and are apparently orbiting each other.

Using Hubble's law, astronomers can immediately gauge a galaxy's distance from its red shift, giving a powerful method for assessing the size and age of the Universe. Some objects have red shifts which indicate

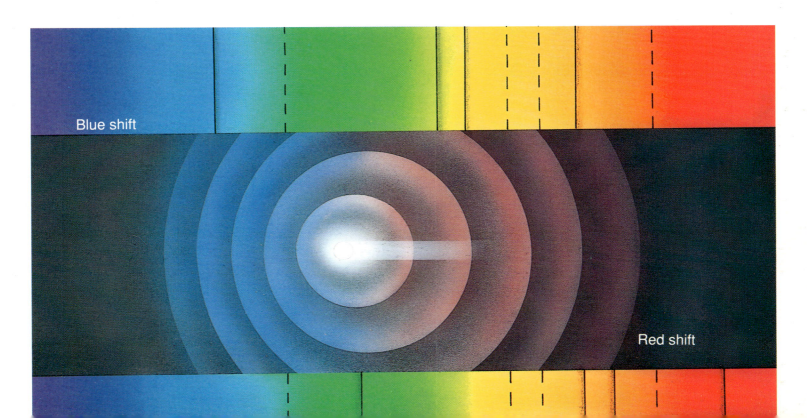

Blue shift

Red shift

they are receding from us at large fractions of the speed of light, placing them so far off in the Universe that the light we see from them today started on its journey before the Earth was born. In this sense large telescopes are time machines, for they allow us to look far back into the past to see the Universe as it appeared when it was much younger.

Although all galaxies outside the Local Group seem to be fleeing from us at ever-increasing speeds, we should not conclude that we are at the centre of the Universe. Since the space between all clusters is stretching like a rubber sheet, observers in any other galaxy would see the same headlong flight from their home cluster. No cluster of galaxies is stationary with others moving away from it – all are moving away from one another. There is nothing unique about our place in space or, indeed, any other place. The Universe has its own curious geometry that denies the existence of either a centre or an edge.

Expansion requires an impulse, and the discovery of the expansion of the Universe led naturally enough to the supposition that it all started in a massive explosion, termed the *Big Bang*. If we could reverse the current expansion, like running a film backwards, the entire Universe would contract until it became jammed together in a hot, superdense mass. At some time in the remote past, everything – all matter, space, and time – was compacted in such a primeval egg. There was nothing outside it, for it *was* the entire Universe.

For some reason unknown, and possibly unknowable, this intensely condensed soup of matter and energy suffered a gigantic explosion, unleashing the expansion of the Universe. That explosion, the Big Bang, marked the creation of the Universe as we know it, and the Universe has been flying apart ever since. The idea of material flying out into empty space is an incorrect one, for empty space itself did not exist before the Big Bang. The galaxies are not moving through space – it is the space

● *Three theories of the origin of the Universe. According to the Big Bang (top), the Universe began with a giant explosion and has been expanding ever since. The rival Steady State theory (centre) said that the Universe has always looked much the same, with new matter being created to fill the space left as the Universe expands. A third view (bottom) maintains that the current expansion will eventually stop and the Universe will shrink to another Big Bang, when the expansion will start all over again.*

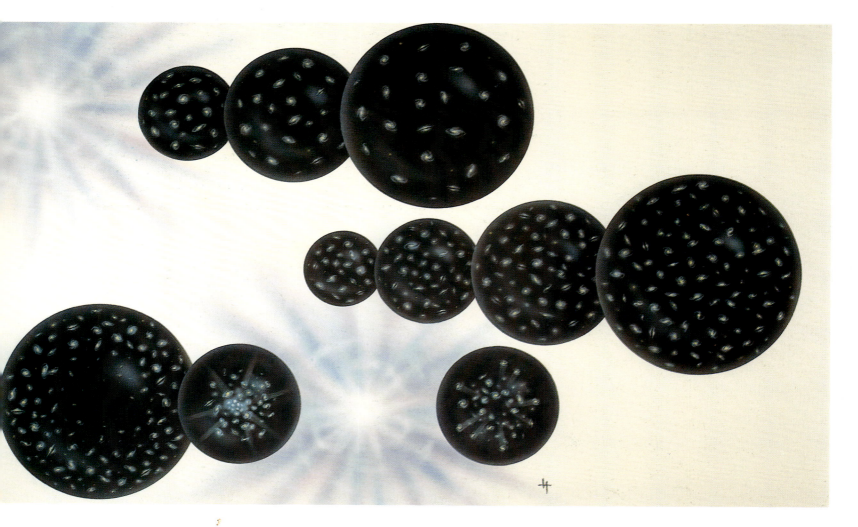

● *Quasars were named because of their quasi-stellar appearance. This (centre) is the brightest of them, 3C 273, in the constellation Virgo, the first to have its red shift measured. It is emitting a jet of gas 150,000 light years long. A similar jet extends from the elliptical galaxy M87.*

between them that is getting bigger. We can now see why there is no centre to the Universe, no single point where the Big Bang happened, because it happened *everywhere*. The entire Universe exploded, not just one part of it.

When did this momentous event take place? That depends on the rate at which the Universe is expanding, as expressed by the disputed value of Hubble's constant. If the faster rates of expansion are correct, the Universe would have taken about 10 billion (thousand million) years to reach its present size, which would make it younger than the oldest known stars. However, taking the average or slowest figures for the expansion gives an age of 15 or 20 billion years, comfortably greater than the age of the oldest stars. Modern astronomy has therefore been able to date the Creation at around 15 billion years ago.

The Big Bang theory was challenged for a while by the *Steady State* theory, proposed in 1948. This did away with Creation by postulating that matter is continually created in space, filling the volume left as the Universe expands. On this theory the Universe would, broadly speaking, have looked pretty much the same 10 or 20 billion years ago as it does today, and would continue to look the same at any time in the future. In fact, the Steady State theory was based on the assumption that the Universe has always existed and always will. Two rapid blows were to discredit the Steady State theory, and at the same time provide new support for the Big Bang.

In the 1960s astronomers were engaged in attempting to identify the optical counterparts of various celestial radio sources. These sources had no names, only numbers prefixed by the designation 3C given to them in the third catalogue compiled by radio astronomers at Cambridge, England. The best telescope for the job was the world's largest, the 5-metre (200-inch) on Palomar Mountain, opened in 1948 some 200 km (125 miles) down the California coast from the venerable Mount Wilson instrument. The technique was to photograph the object to see what it looked like, then take its spectrum to find out more about it.

Several radio sources were quickly identified as galaxies, but others were much more enigmatic: they looked like stars but their spectra initially defied interpretation. Then, in 1963, Palomar astronomer Maarten Schmidt found that one such object, 3C 273 in the constellation Virgo, had a huge red shift that placed it far off in the Universe, about 2000 million light years away. It was not a star at all, but a strange galaxy far brighter than any normal galaxy, and it even sported a jet of ejected gas like the one emanating from the giant elliptical galaxy M87. Once the spectrum of 3C 273 had been interpreted, the others rapidly fell into place. They were called *quasars* because of their quasi-stellar appearance.

Today, thousands of quasars are known, the most distant of them with red shifts that show they are receding at over 90 per cent of the speed of light, placing them as much as 15 billion light years away (depending on the value of Hubble's constant). Whatever their actual distance, the fact is that we are seeing them as they appeared long ago, for when we look out into space we look back in time. There are no quasars around today, but they become much more common the farther away we look. Their very existence proves that the Universe looked different in the past, in contradiction of the Steady State theory. This was the first of the two fatal blows to that theory.

The second came from an unusual direction – the Bell Telephone Laboratories at Holmdel, New Jersey. There, while calibrating the sensitivity of a horn-shaped antenna like a massive ear-trumpet built for experiments with communications satellites, Arno Penzias and Robert Wilson detected a faint but persistent hiss that at first they could not account for. Only when they learned from theorists that the Universe should be pervaded by a slight warmth left over from the Big Bang did they realize the significance of what they had found. It was a whisper of energy, muted by the expansion of the Universe, from the primeval explosion itself. Astronomers term it the *cosmic background radiation*. Its discovery, announced in 1965 and subsequently confirmed by other researchers, earned Penzias and Wilson the Nobel Prize for Physics in 1978.

Understanding what went on in the Big Bang and shortly afterwards is one of the greatest challenges facing modern science. It is believed that all the hydrogen and virtually all the helium in the Universe today were synthesized in the Big Bang fireball. All other elements were formed later by nuclear fusion reactions inside stars (see p. 45).

A few hundred thousand years after the Big Bang, the hydrogen and helium in the Universe had cooled to about 3000°C (the surface temperature of a red giant star) and had thinned out sufficiently to become transparent. The heat emitted at that time, quenched to −270°C by the subsequent expansion of the Universe, is what Penzias and Wilson detected as the cosmic background radiation. Thanks to the Big Bang, space is not entirely cold, even though it is just a scant 3°C above absolute zero, the coldest temperature physically possible (absolute zero is −273°C).

As the Universe expanded and cooled over the millions of years following the Big Bang, giant clouds of hydrogen and helium gas condensed to form galaxies. Details of the process are still vague, but it is thought that elliptical galaxies were spawned by denser clouds of gas in which star formation was rapid, using up most of the available gas and dust. Spirals formed from less dense clouds, the stars of the halo (including globular clusters) and central bulge being produced early on, and the remaining gas and dust settling into a disk where star formation continues today.

But what of quasars? To be visible at such immense distances, they must be vastly more luminous than normal galaxies. Yet their starlike appearance, and the fact that they can vary markedly in brightness over a few months or less, shows that they are much smaller than galaxies. In fact, a typical quasar emits as much light as hundreds of Milky Ways from a region only a few times the size of the Solar System. Although the first quasars were discovered because they were radio sources, it turns out that the great majority of them are not radio sources at all.

Almost certainly, quasars are the ultra-bright centres of young galaxies seen as they appeared not long after the Big Bang, 15 billion or

● *NGC 1566 in Dorado is a Seyfert galaxy, one of a class of spirals with unusually bright centres. They are believed to be related to quasars.*

so years ago. Quasars have much in common with other denizens of deep space known collectively as active galaxies. One category, the *Seyfert galaxies*, are spirals with unusually bright centres that contain clouds of hot, fast-moving gas. They are named after the American astronomer Carl Seyfert who first described them in 1943, although their significance was not realized at the time. The appearance of Seyferts marks them out as scaled-down quasars. Indeed, large telescopes show that certain of the nearer quasars possess fuzzy fringes, evidence of a surrounding galaxy.

Another type of active galaxy is the *BL Lac objects*, named after BL Lacertae, which was originally classified as a strange variable star but which is now known to lie about a billion light years away. They are distinguishable from quasars by the bland nature of their spectra. BL Lac objects seen to be the bright centres of elliptical galaxies, whereas the radio-quiet quasars at least are thought to reside in spirals. Related species are the *N galaxies* (named because of their bright nuclei), and *radio galaxies* (which appear to be emitting radio energy) such as Centaurus A.

This confusing array of labels, a legacy of the different way in which each type of object was discovered, hides an underlying unity. All active galaxies are believed to derive their energy from an immense black hole at their centre with the mass of millions of Suns. The seeds of such gigantic black holes were sown by ancient supernovae, which produced modest holes that grew by merging with others and by swallowing gas from their surroundings. Probably all galaxies are host to central black holes of some sort, but in the case of our own Galaxy the hole never grew large enough to spark off quasar-like activity.

The tidal force of a massive black hole pulls stars apart and gathers the material into a surrounding disk of intensely hot gas a few light weeks in diameter – the size of the brilliant heart of an active galaxy or quasar. Fluctuations in brightness occur as the gas supply changes, and under certain circumstances gas is shot out along the rotation axis of the disk to produce visible jets like those in M87 and 3C 273, and clouds of radio-emitting gas around radio galaxies which can extend for millions of light years.

● *Albert Einstein, originator of the theory of relativity.*

THE TWINS PARADOX

According to Einstein's theory of relativity, the speed of light is the fastest speed in the Universe. This speed limit has some odd consequences. Imagine that you have an identical twin who remains on Earth while you go on a high-speed spaceship ride to a nearby star and back. As your spaceship edged closer to the speed of light, time on board would slow down, an effect known as *time dilation*. It would affect not just the ticking of clocks, but all temporal processes such as your heartbeat and ageing. Relativity predicts that, in step with this slowing of time, the spaceship and everything on board would shrink in length and increase in mass. At the speed of light itself, mass would become infinite, length would become zero, and time would stand still, which is why it is impossible ever to reach the speed of light.

None of these processes, though, would be apparent to you on the spaceship, and everything would return to its Earth value once you decelerated. Only when you returned home would you realize that anything unusual had happened, for your identical twin would now be much older than you. This is the so-called Twins Paradox. Although the prediction seems to defy common sense, we can see its effects at work in the case of cosmic rays, atomic particles which arrive at the Earth at high speed. Their masses and lifetimes are increased, just as predicted by Einstein. Time dilation offers a form of one-way time travel into the future. If we could travel sufficiently close to the speed of light we could cross the Galaxy in a human lifetime, while on Earth 100,000 years would pass.

The black hole continues to disrupt stars, feed on the products, and emit radiation equivalent to the energy of entire galaxies for perhaps a billion years until it runs out of stars to supply it. In a number of Seyferts, radio galaxies, and quasars, the activity is stoked by inter-actions or mergers with neighbouring galaxies which rain fresh fuel onto the central black hole. Such encounters would have been more common in the past when the Universe was smaller. 'Dead' quasars in the form of massive, quiescent black holes must now lurk at the heart of many nearby galaxies.

Astronomers now seem to have established what happened to the Universe in the past, but what of its future? If the Universe is dense enough, the current expansion will eventually slow down and stop, to be replaced by a contraction back to what is sometimes called the Big Crunch, followed by another Big Bang. In this scenario, the Universe may oscillate endlessly between expansion and contraction, a new Universe emerging from each Big Bang.

STAR MAPS

On the following pages is an atlas of the entire sky divided into ten sections. Maps 1 and 2 show the north polar region of the sky, Maps 9 and 10 the south polar region, and Maps 3 to 8 the equatorial region. Dot sizes indicate the brightness of the stars, down to fifth magnitude. Special symbols are used to distinguish double and multiple stars, variable stars, and for other objects such as star clusters, nebulae, and galaxies labelled with their Messier (M) numbers where they have them, or their NGC numbers (without prefix) or IC numbers (prefix I.). The light blue band is the Milky Way. (For more about the naming of celestial objects, see the boxes on pages 10 and 25)

The coordinates on the maps are *right ascension* and *declination*, the celestial equivalents of terrestrial longitude and latitude. Declination, like latitude, is measured from 0° at the celestial equator to 90° at the poles. Right ascension is measured not in degrees but in hours, from 0 to 24, reflecting the daily rotation of the Earth.

The dashed line on the maps labelled *ecliptic* represents the Sun's yearly path against the star background. Actually, of course, the Sun's movement is only illusory – it is really the Earth that is moving, completing one orbit every year. Because the Earth's axis is tilted, the ecliptic forms an angle with the celestial equator, about 23½°. Right ascension is measured from the point where the Sun's path crosses the celestial equator into the northern hemisphere. The Sun reaches this point in late March each year at the *spring equinox*.

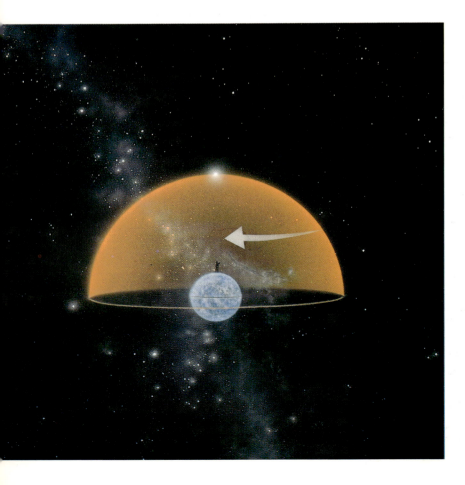

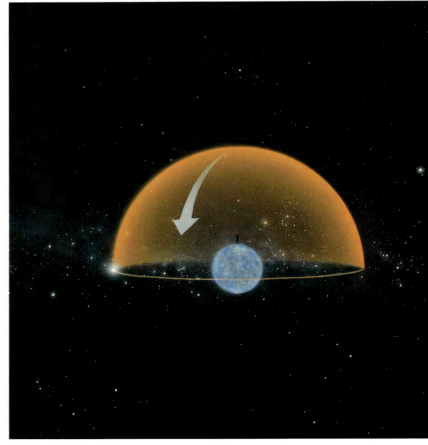

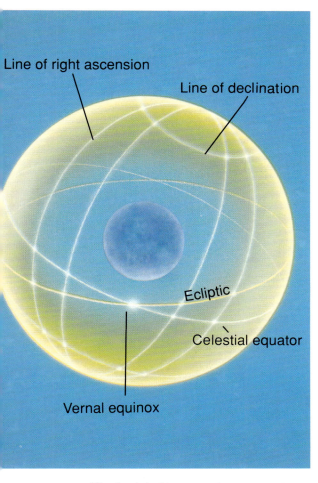

Line of right ascension

Line of declination

Ecliptic

Celestial equator

Vernal equinox

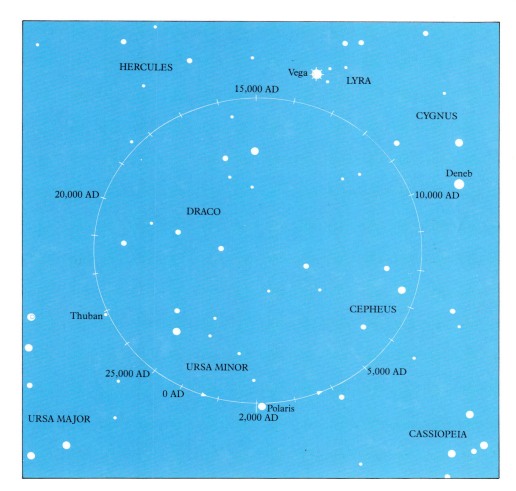

● *Precession causes the Earth's north pole to trace out a circle on the celestial sphere every 26,000 years.*

● *All celestial objects can be imagined to lie on the imaginary celestial sphere surrounding the Earth. The celestial poles lie directly above the poles of the Earth, and the celestial equator lies directly above the Earth's equator. The Sun's annual path around the sky, the ecliptic, cuts the celestial equator at a point called the spring (or vernal) equinox. Right ascension is measured from this point.*

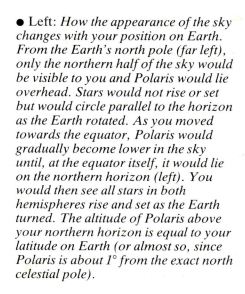

● *Left: How the appearance of the sky changes with your position on Earth. From the Earth's north pole (far left), only the northern half of the sky would be visible to you and Polaris would lie overhead. Stars would not rise or set but would circle parallel to the horizon as the Earth rotated. As you moved towards the equator, Polaris would gradually become lower in the sky until, at the equator itself, it would lie on the northern horizon (left). You would then see all stars in both hemispheres rise and set as the Earth turned. The altitude of Polaris above your northern horizon is equal to your latitude on Earth (or almost so, since Polaris is about 1° from the exact north celestial pole).*

The celestial poles lie directly overhead at the Earth's poles. It so happens that a moderately bright star, Polaris, lies near the north celestial pole, but there is no equivalent pole star in the southern hemisphere. However, because of a slow wobble of the Earth in space, called *precession*, the position of the celestial poles against the stars changes steadily, completing one full circle every 26,000 years. Star positions change by about 1° every 70 years because of precession. The star positions on the maps here are plotted for the year 2000, although the differences will be negligible for several decades after that date. One consequence of precession is that Polaris is only temporarily close to the celestial pole.

Twelve additional charts on pages 178 to 189 show a strip of sky arching overhead from the northern to the southern horizon as it appears each month around 10 pm in mid-month (11 pm when daylight saving time is in operation). These charts will enable you to locate various stars and constellations at a convenient time of night. Fewer stars are plotted than on the all-sky charts – down to fourth magnitude instead of fifth.

Horizon lines are indicated for latitudes ranging from the equator to 60°N. Note that the altitude of Polaris above your northern horizon is almost equal to your latitude on Earth, since Polaris is currently only 1° away from the celestial pole.

If you are observing at a date and time different from those shown on the monthly charts, you will have to change maps. For every two hours *later* than 10 pm, go one map *forward*; for every two hours *earlier* than 10 pm, turn *back* one map. For example, if you are observing at midnight in mid-January turn to the February map; if at 8pm in mid-January, use the December map.

THE 88 CONSTELLATIONS

NAME	GENITIVE	ABBR.	AREA (square degrees)	ORDER OF SIZE	MAP(S)
Andromeda	Andromedae	And	722	19	8, 2, 3
Antlia	Antliae	Ant	239	62	6
Apus	Apodis	Aps	206	67	9
Aquarius	Aquarii	Aqr	980	10	3
Aquila	Aquilae	Aql	652	22	4, 3
Ara	Arae	Ara	237	63	9
Aries	Arietis	Ari	441	39	8
Auriga	Aurigae	Aur	657	21	7, 2
Boötes	Boötis	Boo	907	13	5, 1
Caelum	Caeli	Cae	125	81	7
Camelopardalis	Camelopardalis	Cam	757	18	2, 1
Cancer	Cancri	Cnc	506	31	6
Canes Venatici	Canum Venaticorum	CVn	465	38	5, 1
Canis Major	Canis Majoris	CMa	380	43	7
Canis Minor	Canis Minoris	CMi	183	71	7
Capricornus	Capricorni	Cap	414	40	3
Carina	Carinae	Car	494	34	9
Cassiopeia	Cassiopeiae	Cas	598	25	2
Centaurus	Centauri	Cen	1060	9	9, 5, 6
Cepheus	Cephei	Cep	588	27	2
Cetus	Ceti	Cet	1231	4	8
Chamaeleon	Chamaeleontis	Cha	132	79	9
Circinus	Circini	Cir	93	85	9
Columba	Columbae	Col	270	54	7
Coma Berenices	Comae Berenices	Com	386	42	5
Corona Australis	Coronae Australis	CrA	128	80	4
Corona Borealis	Coronae Borealis	CrB	179	73	5
Corvus	Corvi	Crv	184	70	5
Crater	Crateris	Crt	282	53	6
Crux	Crucis	Cru	68	88	9
Cygnus	Cygni	Cyg	804	16	2, 3, 4
Delphinus	Delphini	Del	189	69	3
Dorado	Doradus	Dor	179	72	9, 10
Draco	Draconis	Dra	1083	8	1, 2
Equuleus	Equulei	Equ	72	87	3
Eridanus	Eridani	Eri	1138	6	7, 8, 10
Fornax	Fornacis	For	398	41	8
Gemini	Geminorum	Gem	514	30	7
Grus	Gruis	Gru	366	45	1, 3
Hercules	Herculis	Her	1225	5	4, 1, 2
Horologium	Horologii	Hor	249	58	10
Hydra	Hydrae	Hya	1303	1	6, 5
Hydrus	Hydri	Hyi	243	61	10
Indus	Indi	Ind	294	49	10
Lacerta	Lacertae	Lac	201	68	2, 3
Leo	Leonis	Leo	947	12	6
Leo Minor	Leonis Minoris	LMi	232	64	6
Lepus	Leporis	Lep	290	51	7
Libra	Librae	Lib	538	29	5
Lupus	Lupi	Lup	334	46	9, 5
Lynx	Lyncis	Lyn	545	28	1
Lyra	Lyrae	Lyr	286	52	4
Mensa	Mensae	Men	153	75	9
Microscopium	Microscopii	Mic	210	66	3
Monoceros	Monocerotis	Mon	482	35	7
Musca	Muscae	Mus	138	77	9
Norma	Normae	Nor	165	74	9
Octans	Octantis	Oct	291	50	9
Ophiuchus	Ophiuchi	Oph	948	11	4
Orion	Orionis	Ori	594	26	7
Pavo	Pavonis	Pav	378	44	10,9
Pegasus	Pegasi	Peg	1121	7	3

Perseus	Persei	Per	615	24	2, 7
Phoenix	Phoenicis	Phe	469	37	10, 3
Pictor	Pictoris	Pic	247	59	9
Pisces	Piscium	Psc	889	14	8, 3
Piscis Austrinus	Piscis Austrini	PsA	245	60	3
Puppis	Puppis	Pup	673	20	7, 9
Pyxis	Pyxidis	Pyx	221	65	6
Reticulum	Reticuli	Ret	114	82	10
Sagitta	Sagittae	Sge	80	86	4
Sagittarius	Sagittarii	Sgr	867	15	4
Scorpius	Scorpii	Sco	497	33	4
Sculptor	Sculptoris	Scl	475	36	8, 3
Scutum	Scuti	Sct	109	84	4
Serpens	Serpentis	Ser	637	23	4, 5
Sextans	Sextantis	Sex	314	47	6
Taurus	Tauri	Tau	797	17	7, 8
Telescopium	Telescopii	Tel	252	57	10
Triangulum	Trianguli	Tri	132	78	8
Triangulum Australe	Trianguli Australis	TrA	110	83	9
Tucana	Tucanae	Tuc	295	48	10
Ursa Major	Ursae Majoris	UMa	1280	3	1, 6
Ursa Minor	Ursae Minoris	UMi	256	56	1, 2
Vela	Velorum	Vel	500	32	9
Virgo	Virginis	Vir	1294	2	5
Volans	Volantis	Vol	141	76	9
Vulpecula	Vulpeculae	Vul	268	55	3, 4

TOP TWENTY BRIGHTEST STARS

NAME	STAR	COORDINATES RA h m	Dec. °	MAG.
α Canis Majoris	Sirius	06 45.1	−16 43	−1.46
α Carinae	Canopus	06 24.0	−52 42	−0.72
α Centauri	Rigil Kentaurus	14 39.6	−60 50	−0.27[a]
α Boötis	Arcturus	14 15.7	+19 11	−0.04
α Lyrae	Vega	18 36.9	+38 47	+0.03
α Aurigae	Capella	05 16.7	+46 00	0.08
β Orionis	Rigel	05 14.5	−08 12	0.12
α Canis Minoris	Procyon	07 39.3	+05 13	0.38
α Eridani	Achernar	01 37.7	−57 14	0.46
α Orionis	Betelgeuse	05 55.2	+07 24	0.50[b]
β Centauri	Hadar	14 03.8	−60 22	0.61
α Aquilae	Altair	19 50.8	+08 52	0.77
α Crucis	Acrux	12 26.6	−63 06	0.79[a]
α Tauri	Aldebaran	04 35.9	+16 31	0.85[b]
α Scorpii	Antares	16 29.4	−26 26	0.96[b]
α Virginis	Spica	13 25.2	−11 10	0.98
β Geminorum	Pollux	07 45.3	+28 01	1.14
α Piscis Austrini	Fomalhaut	22 57.6	−29 37	1.16
α Cygni	Deneb	20 41.4	+45 17	1.25
β Crucis	Becrux	12 47.7	−59 41	1.25

[a] Combined magnitude of double star.
[b] Average magnitude of variable star.

Data from the Yale *Bright Star Catalogue*, 4th edition

THE GREEK ALPHABET

α	alpha	ν	nu
β	beta	ξ	xi
γ	gamma	o	omicron
δ	delta	π	pi
ε	epsilon	ρ	rho
ζ	zeta	σ	sigma
η	eta	τ	tau
θ	theta	υ	upsilon
ι	iota	φ	phi
κ	kappa	χ	chi
λ	lambda	ψ	psi
μ	mu	ω	omega

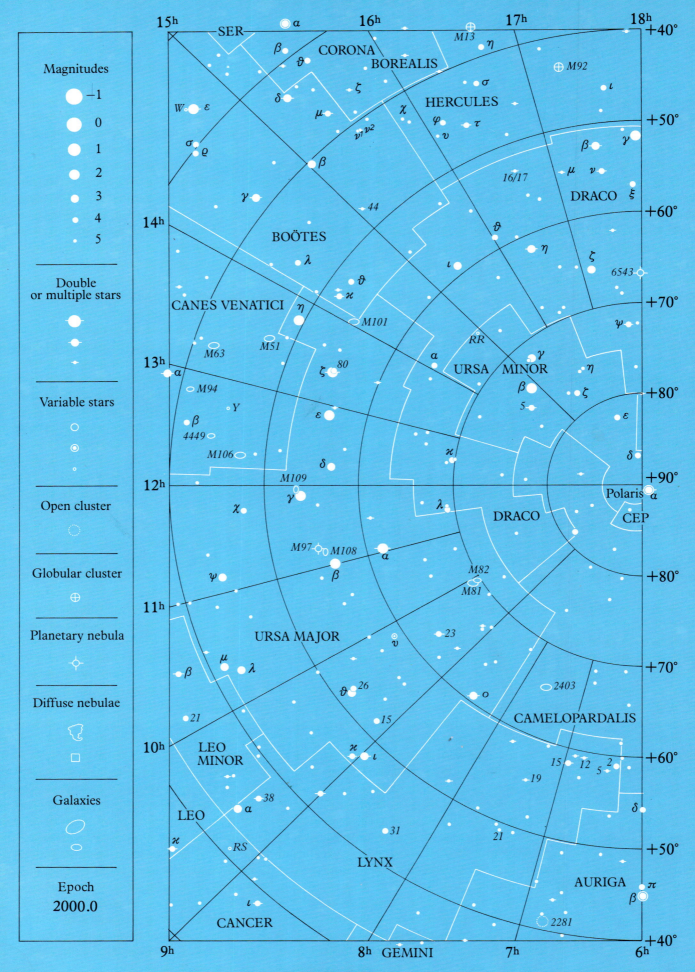

MAP 1

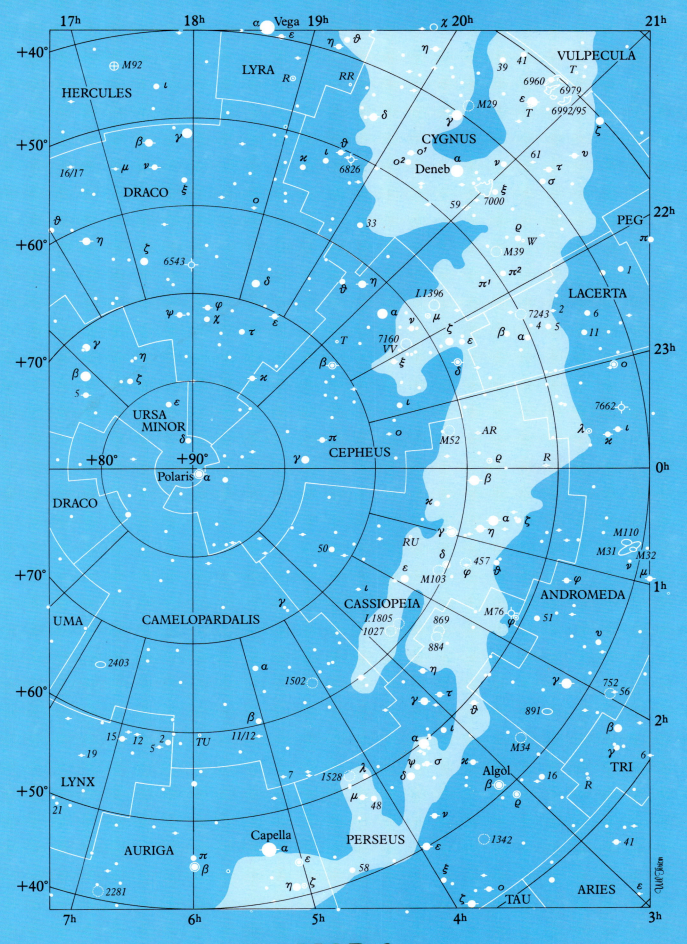

MAP 2

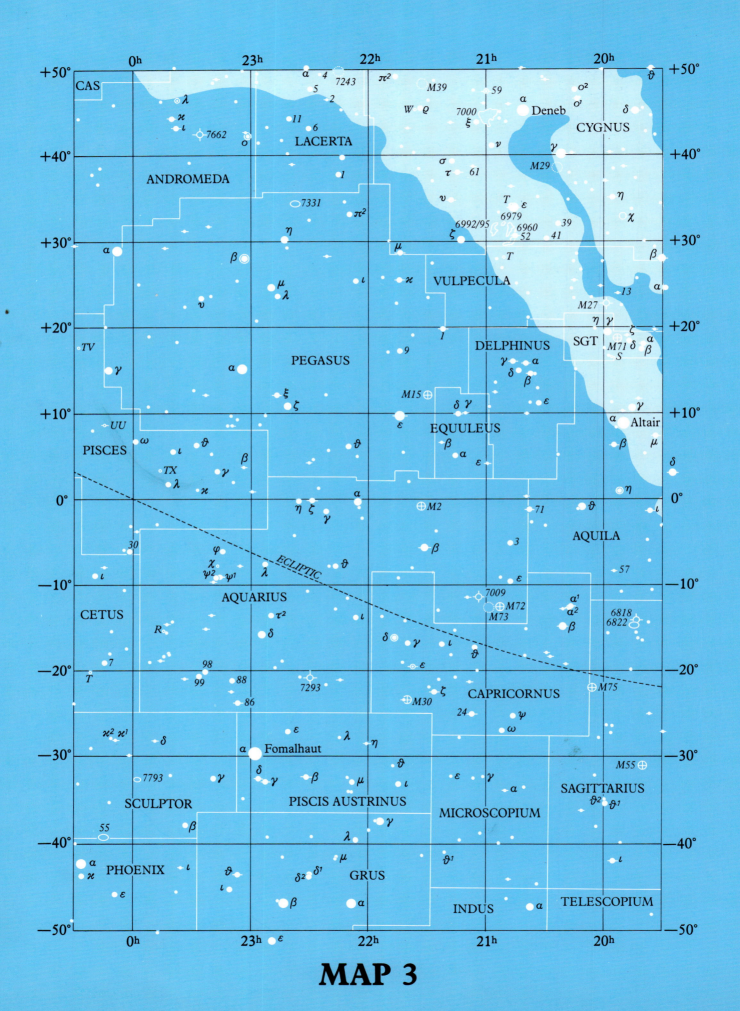

MAP 3

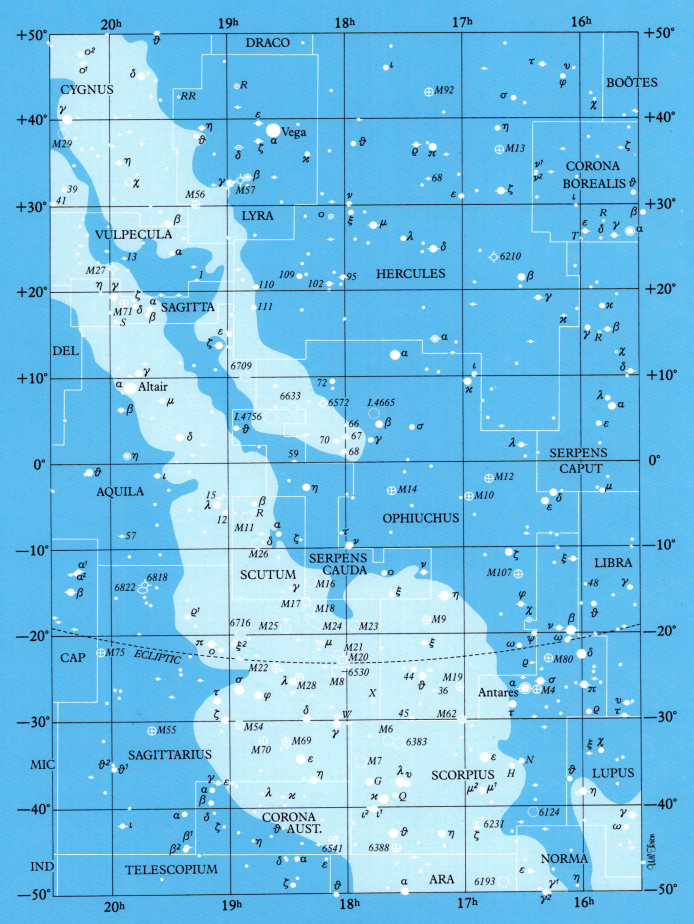

MAP 4

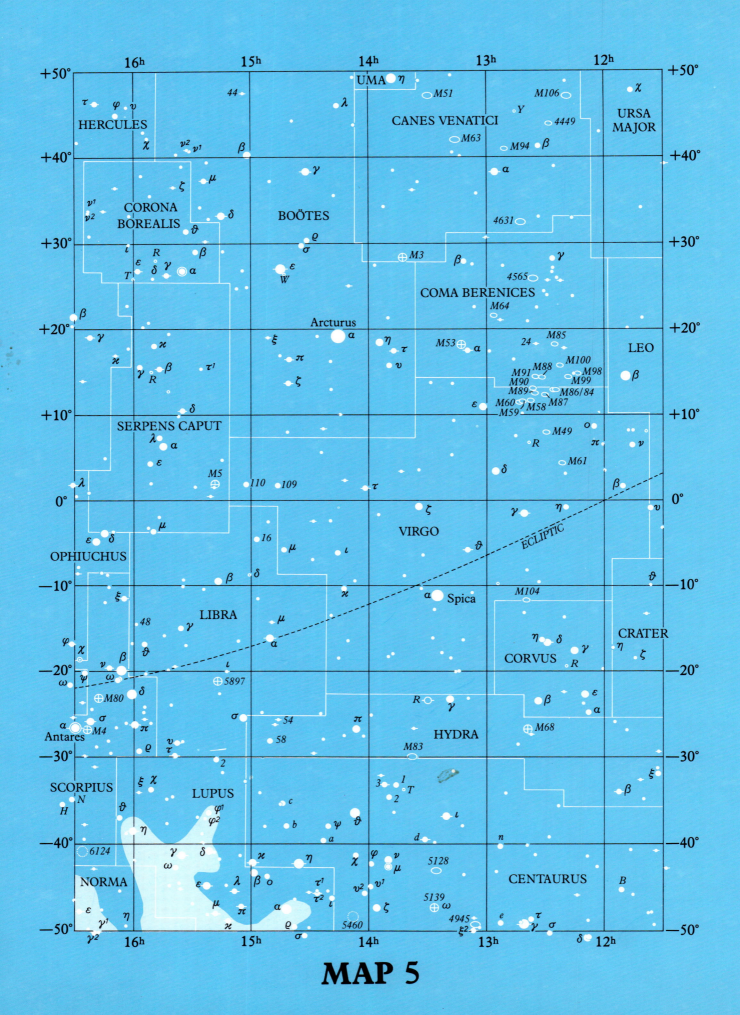

MAP 5

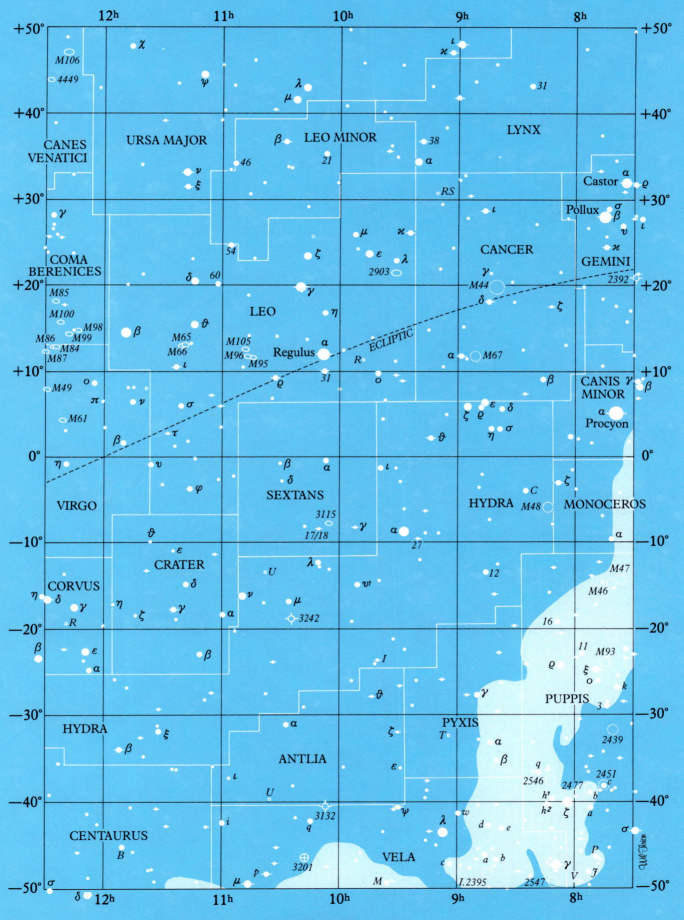

MAP 6

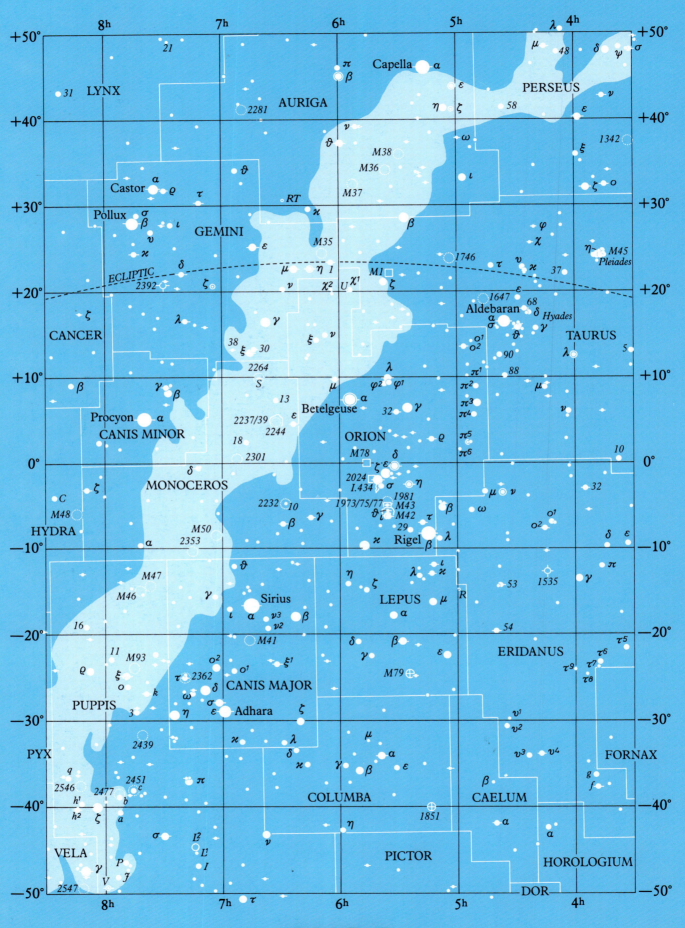

MAP 7

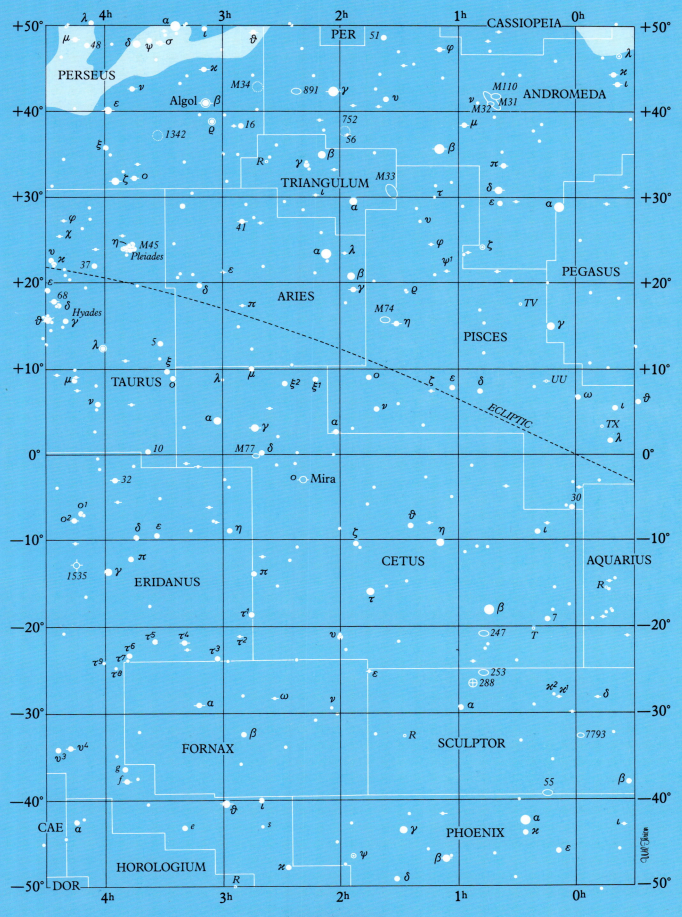

MAP 8

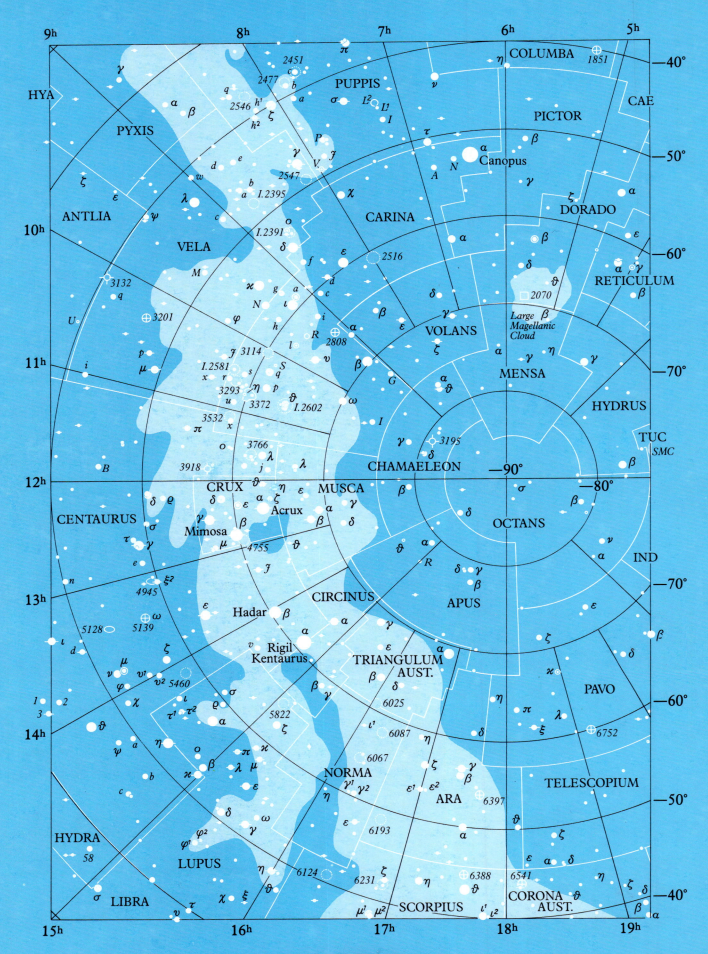

MAP 9

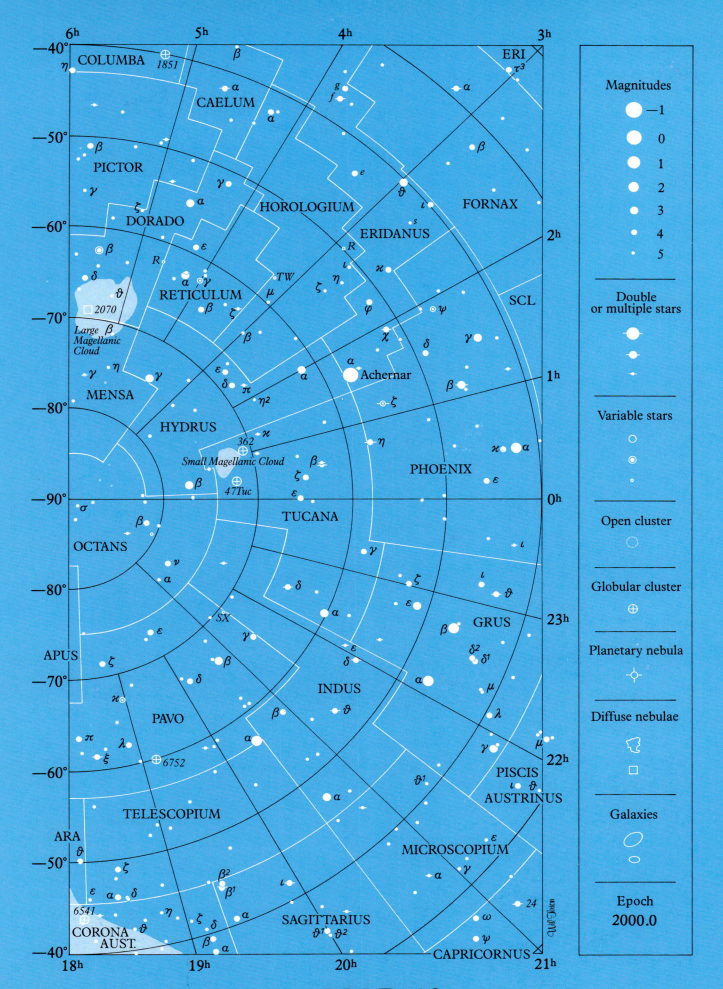

MAP 10

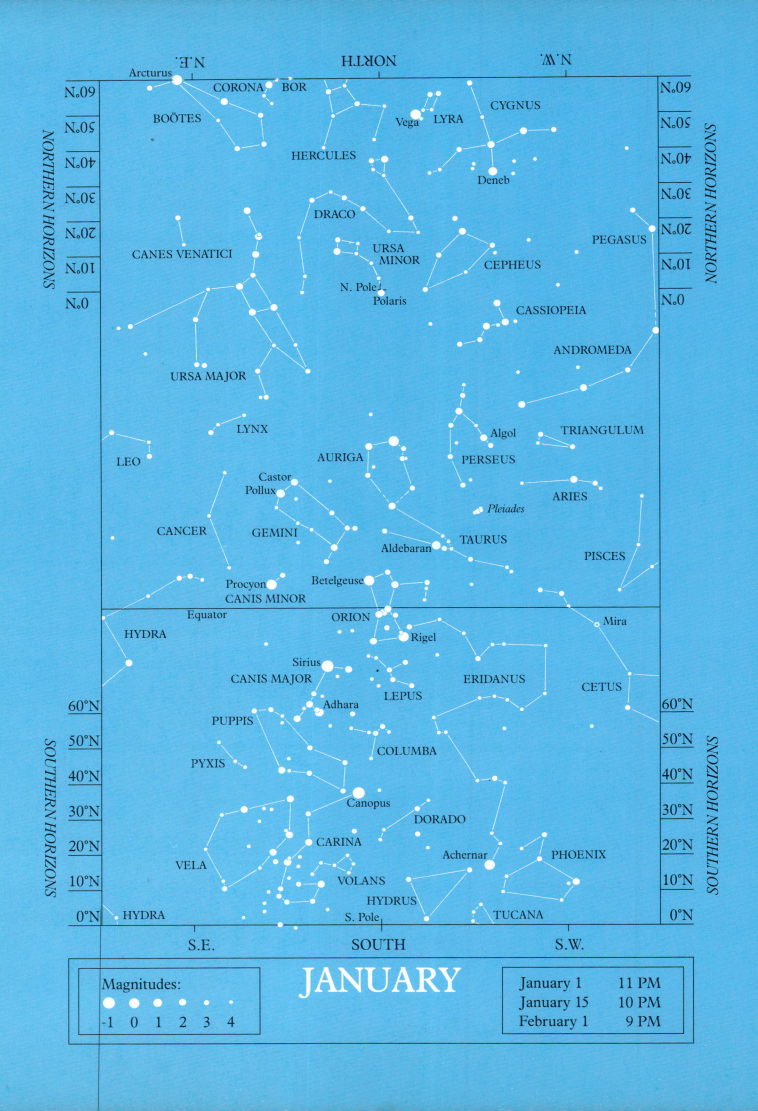

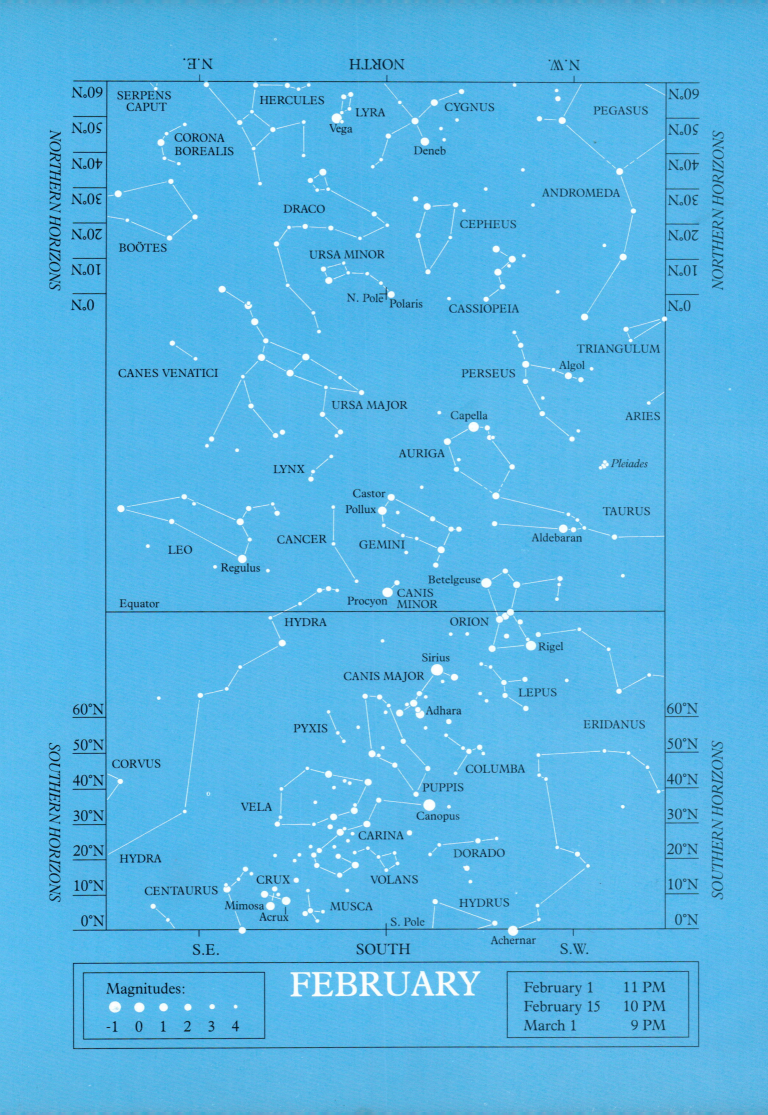

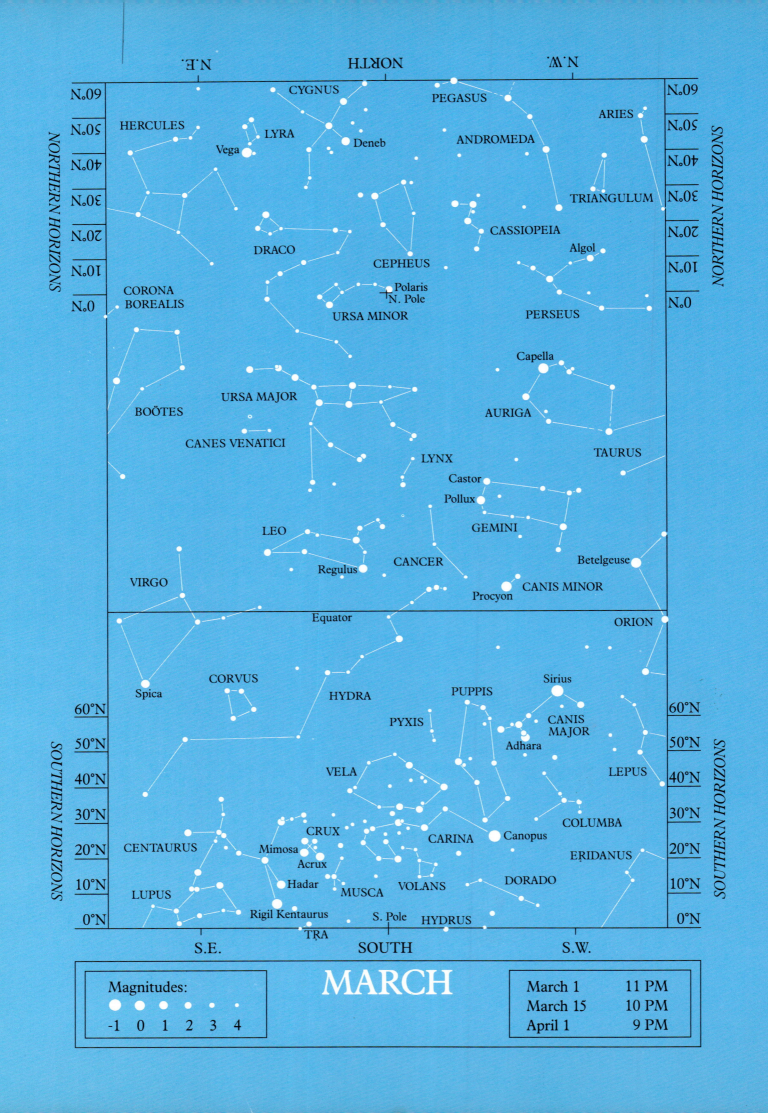

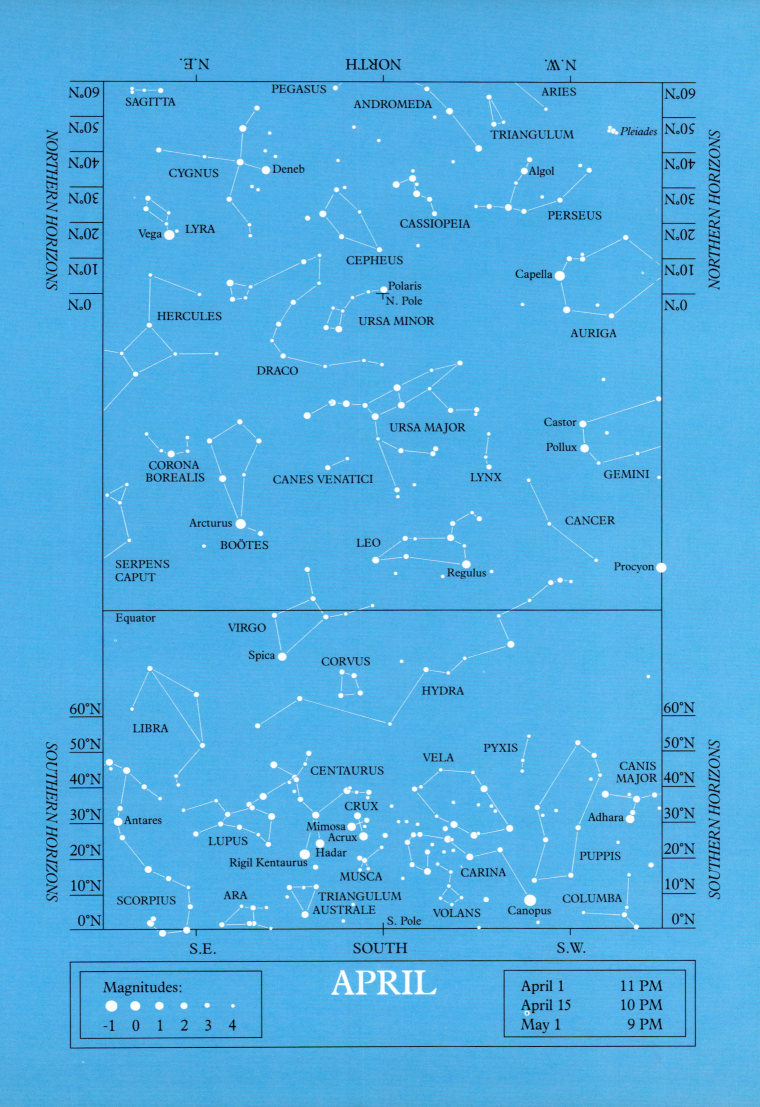

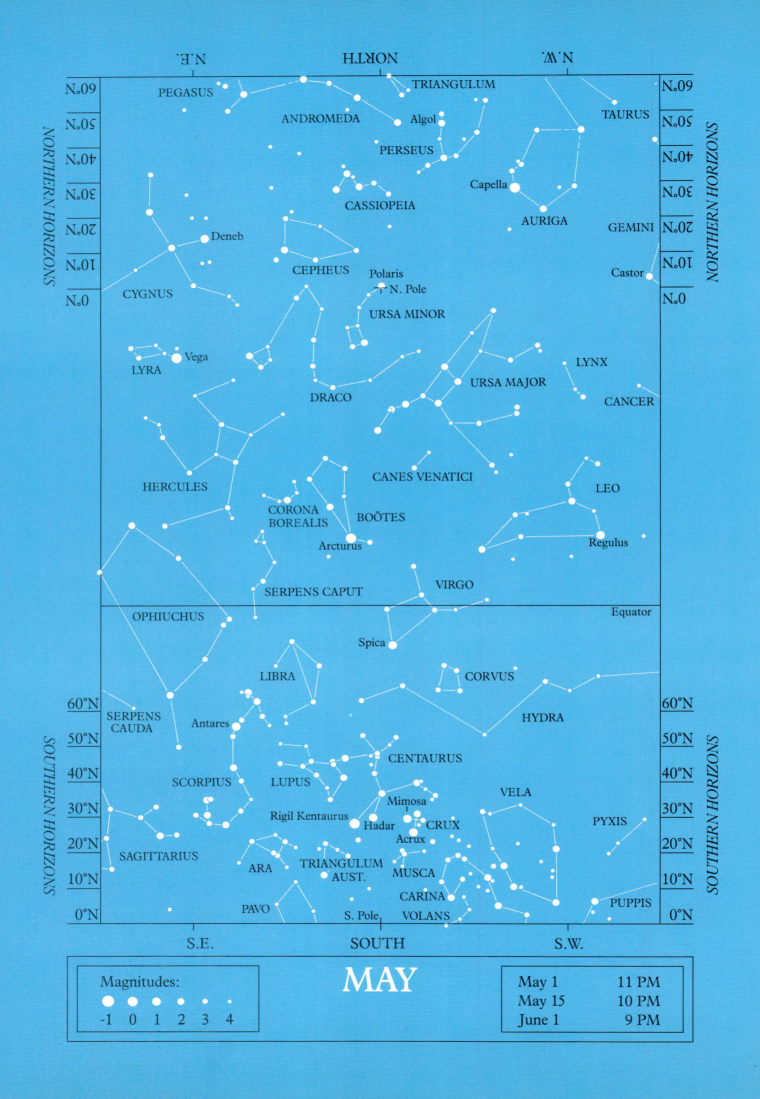

MAY

May 1	11 PM	
May 15	10 PM	
June 1	9 PM	

Magnitudes:
-1 0 1 2 3 4

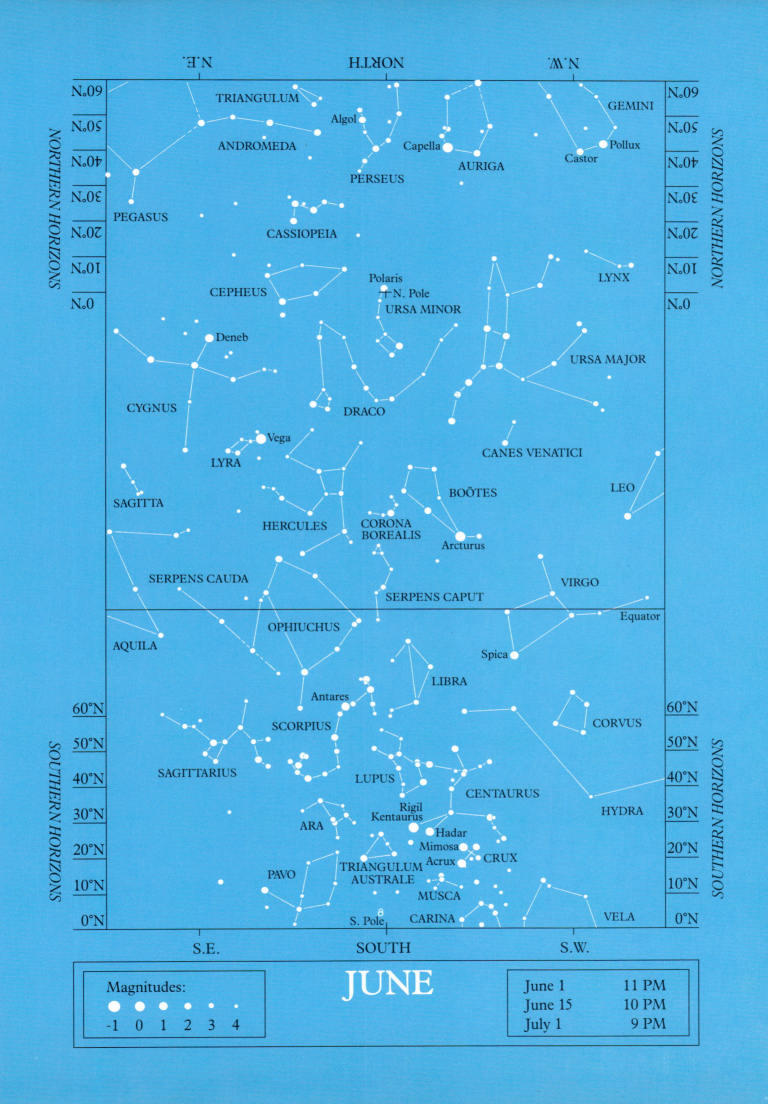

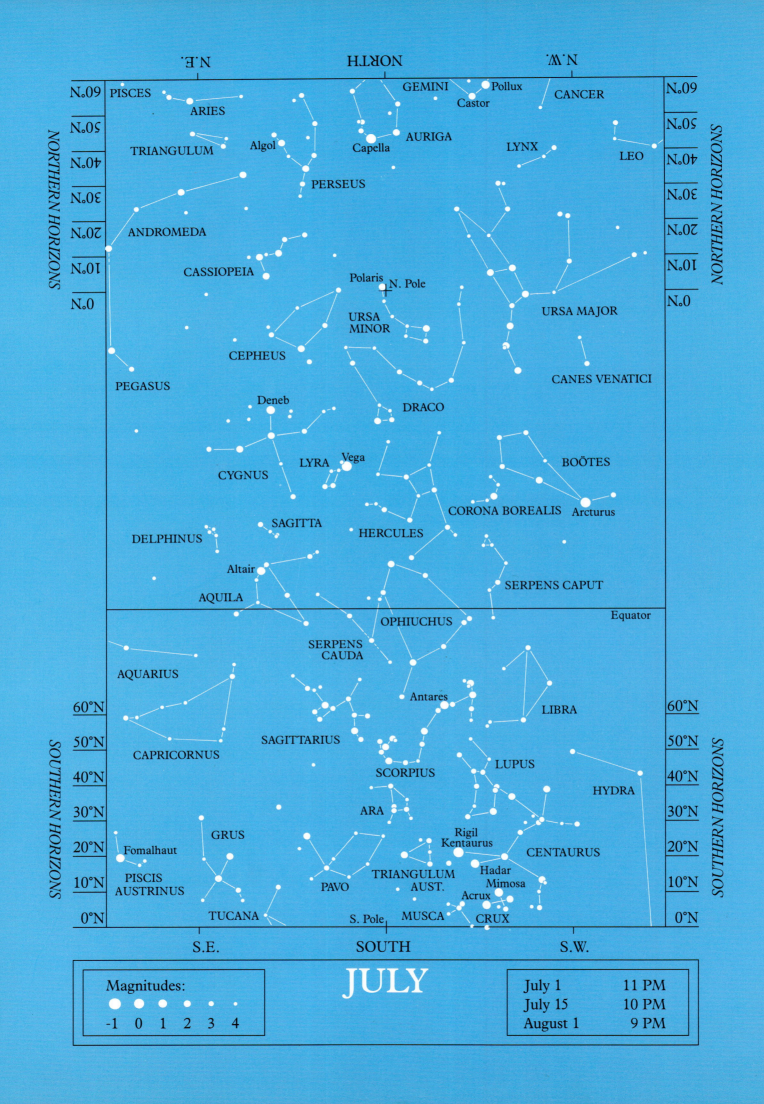

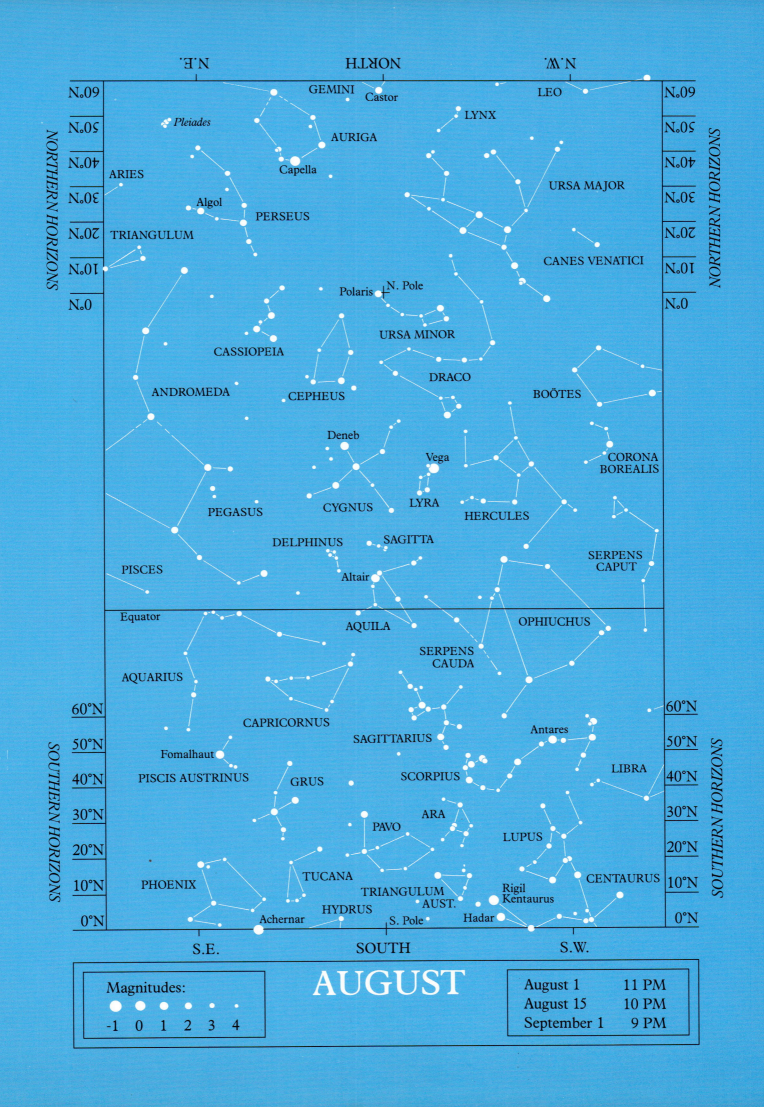

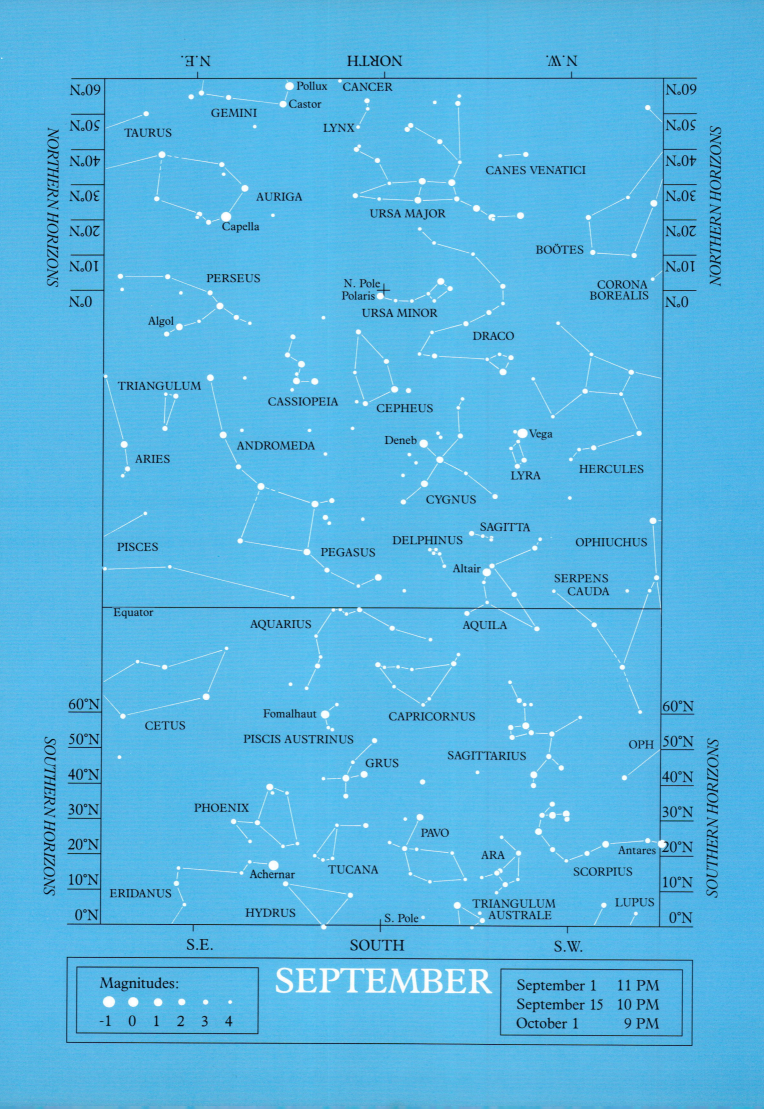

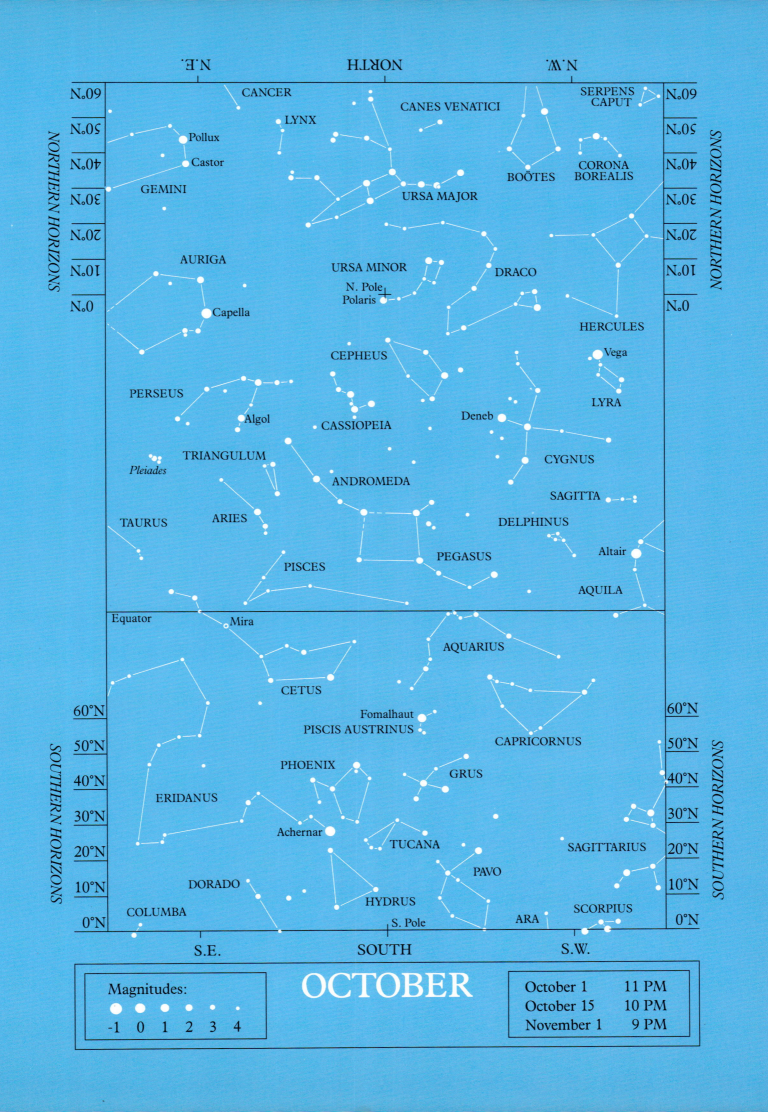

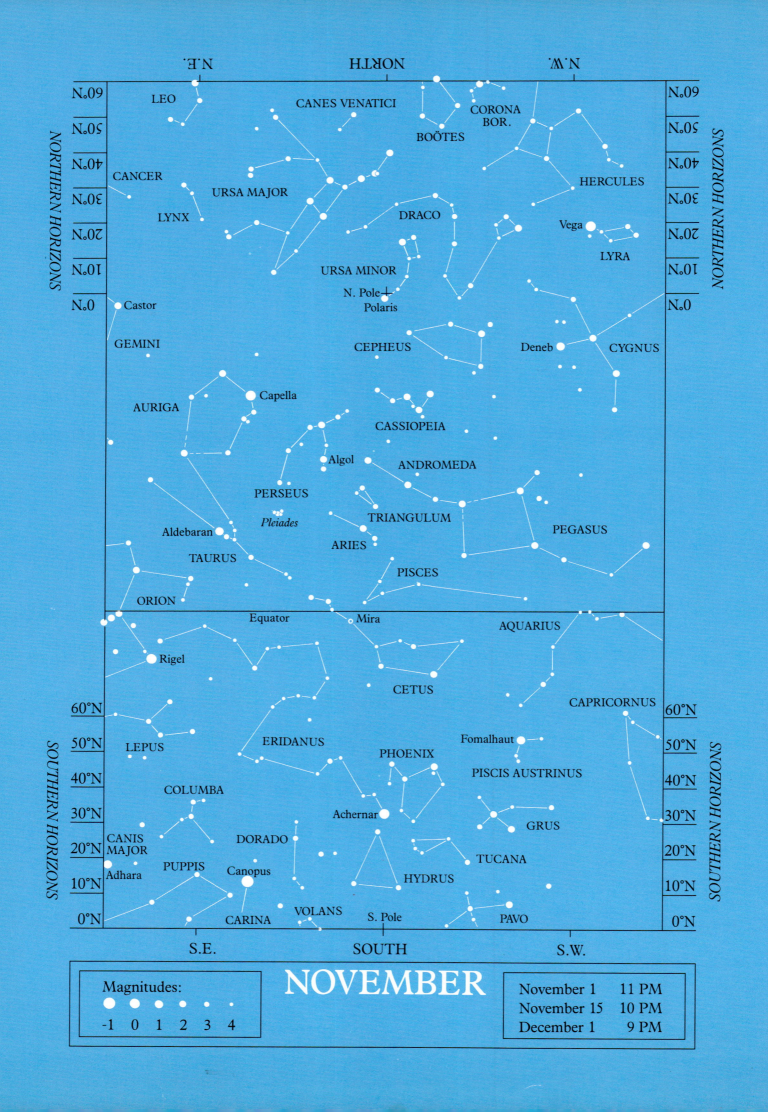

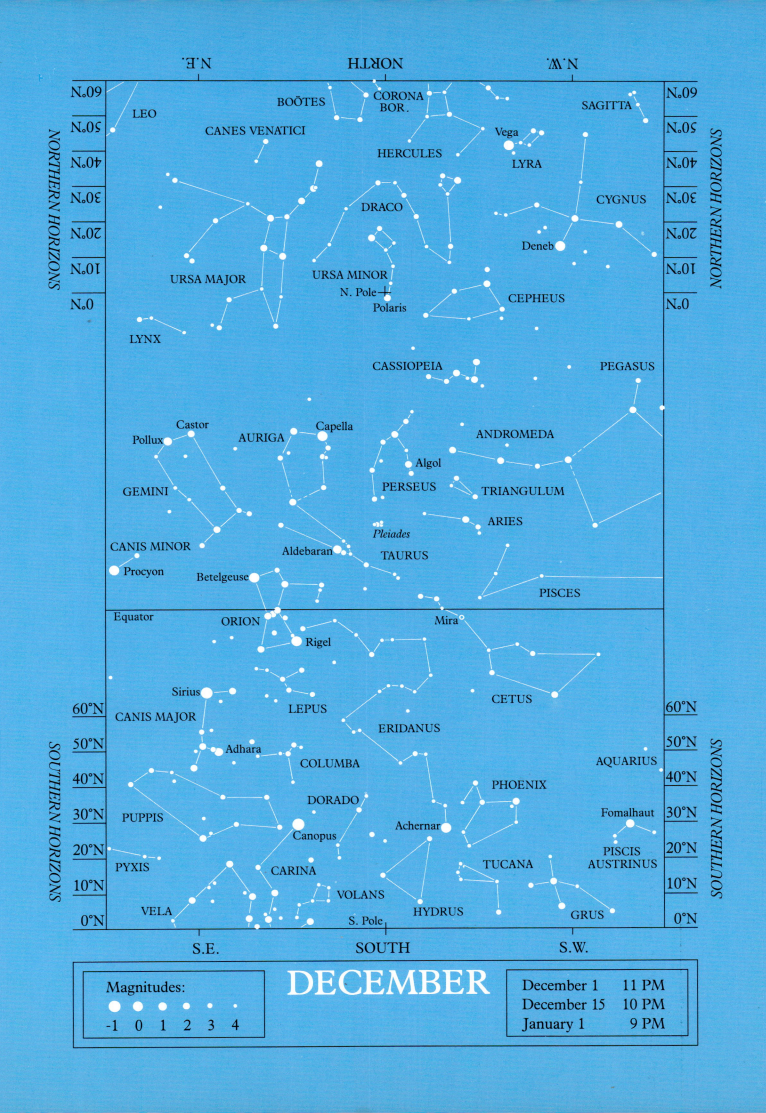

Index

Picture credits